Environments Through Time

A Laboratory Manual in Historical Geology

Second Edition

Robert L. Anstey
Michigan State University
East Lansing, Michigan

Illustrated by:
Terry L. Chase
University of Minnesota
Minneapolis, Minnesota

Burgess Publishing Company
Minneapolis, Minnesota

Printed in the United States of America
ISBN 0-8087-0051-1

0 9 8 7 6 5 4 3 2

Preface

Environments through Time is designed to help beginning geology students with the practical interpretation of sedimentary rocks and fossils and to show them how rocks and fossils are used in determining geologic time and reconstructing the earth's history. Although this manual will generally accompany a course in historical geology, it can also be used in beginning courses in stratigraphy, sedimentary environments, and paleoecology.

The exercises begin with the smallest convenient scale of observation, individual sediment grains, and proceed step by step to larger and larger scales of observation, ending with an analysis of global tectonics. Along the way, students will study dynamic models in stratigraphy and methods of determining geologic time relationships. In the last few exercises, they will be introduced to the concept of simulation models and learn the uses of a few simple ones. Processes that literally take place overnight are studied, as well as those that take eons.

Some of the mathematics problems in Chapter 4 require the use of algebra, but all the other chapters are nonquantitative and use nothing more complicated than graph paper. Although it is easier to understand the exercises when they are taken in order, they can be rearranged to some extent without serious loss of continuity. When time does not permit a full week's treatment of each exercise, several can be combined or assigned as outside reading, as suggested in the instructor's guide.

Improvements in this revised edition include new chapters on geologic age dating, interpretation of the earth's history from geologic maps, and sedimentation and plate tectonics. These three new chapters enlarge the scope of the manual and make it more suitable for historical geology courses. Changing concepts in geology and paleontology also necessitated many alterations in the manual. The chapters on fossil communities and the origin of dolomite have been completely revised. In addition, all the remaining chapters, especially those on invertebrate fossils and patterns of speciation, have been improved and updated. And, finally, several new illustrations prepared by Terry L. Chase contribute to both the usefulness and the appearance of this new edition.

Note to Students

Graph paper, worksheets, and tracing paper for many of the exercises will be found at the back of the book.

Note to Instructors

An instructor's guide accompanies this manual and will be supplied by the publisher upon request. The rocks, fossils, and sediment samples are of the most common types and can be collected in the field in most parts of the United States and Canada. Although field collections are ideal for this manual, commercially available specimens can also be used. Adequate materials for most of the exercises can be obtained from Ward's Natural Science Establishment, Inc. (Rochester, New York, and Monterey, California). All geologic and topographic maps required can be ordered from the United States Geological Survey.

Acknowledgments

Among the many people who have helped us with their contributions and criticisms, we are especially grateful to Mary Davis, Duncan Sibley, Grahame Larson, and Russell Harmon.

Contents

Chapter One

Sedimentology Characteristics of Detrital Sediment Grains

Objectives

1. To understand the simple sedimentary processes of weathering, transportation, and deposition.
2. To determine the effects of different amounts of kinetic energy on the roundness, size, and sorting of sediment grains.
3. To compare and contrast the roundness, size, and sorting of samples taken from different sedimentary environments.

Important Terms

Detrital—Derived from a preexisting rock by weathering and transportation; grains are predominantly of silicate minerals.
Roundness—The absence of angular corners and edges from a sediment grain (Figure 1.1).
Sorting—The uniformity of grain size within a sediment (Figure 1.2).

Introduction

Sedimentation involves the removal (and possible chemical alteration) of a rock grain from its parent material (weathering), its transportation by a fluid medium (air, water, or ice), and its final settling out of that fluid (deposition). According to a physical relationship known as Stokes' law, larger particles will settle out of quiet water first, whereas smaller particles tend to remain in suspension. The higher the water velocity or degree of water agitation, the larger the particle that can stay in suspension. Grain roundness depends upon the number of times that grains come into mutual contact. Different degrees of grain roundness are shown in Figure 1.1. Roundness could reflect either the degree of water agitation or the length of time that the grain has remained in suspension.

Sorting is the absence of grain size variation within a sediment. Samples that vary little in size from grain to grain are well sorted, whereas those with considerable variation are poorly sorted. To observe the size variation in different degrees of sorting, mark the grains which have the largest and

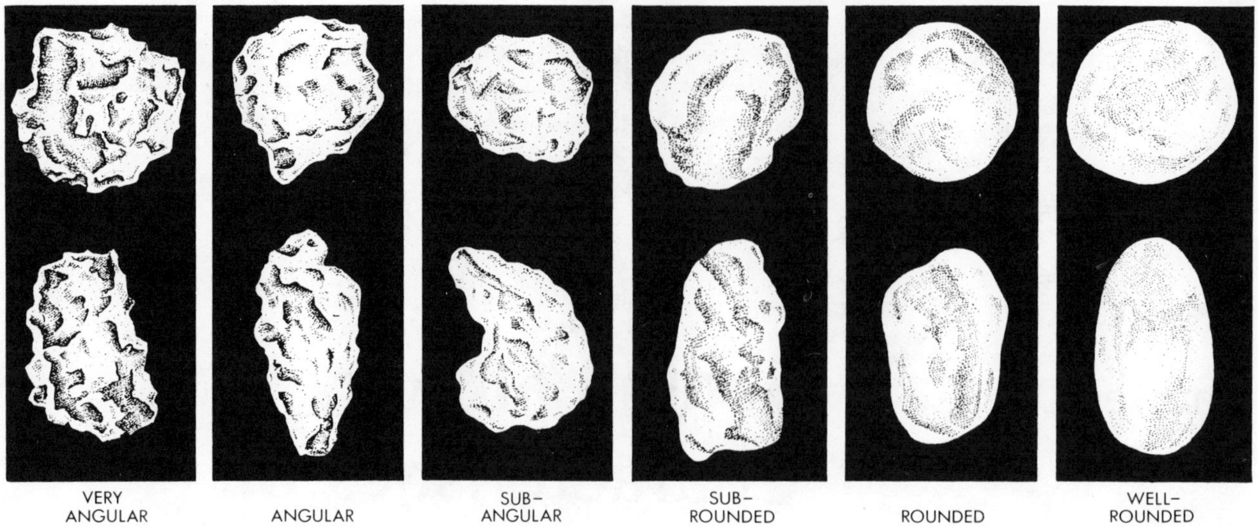

Figure 1.1. Grain roundness images.

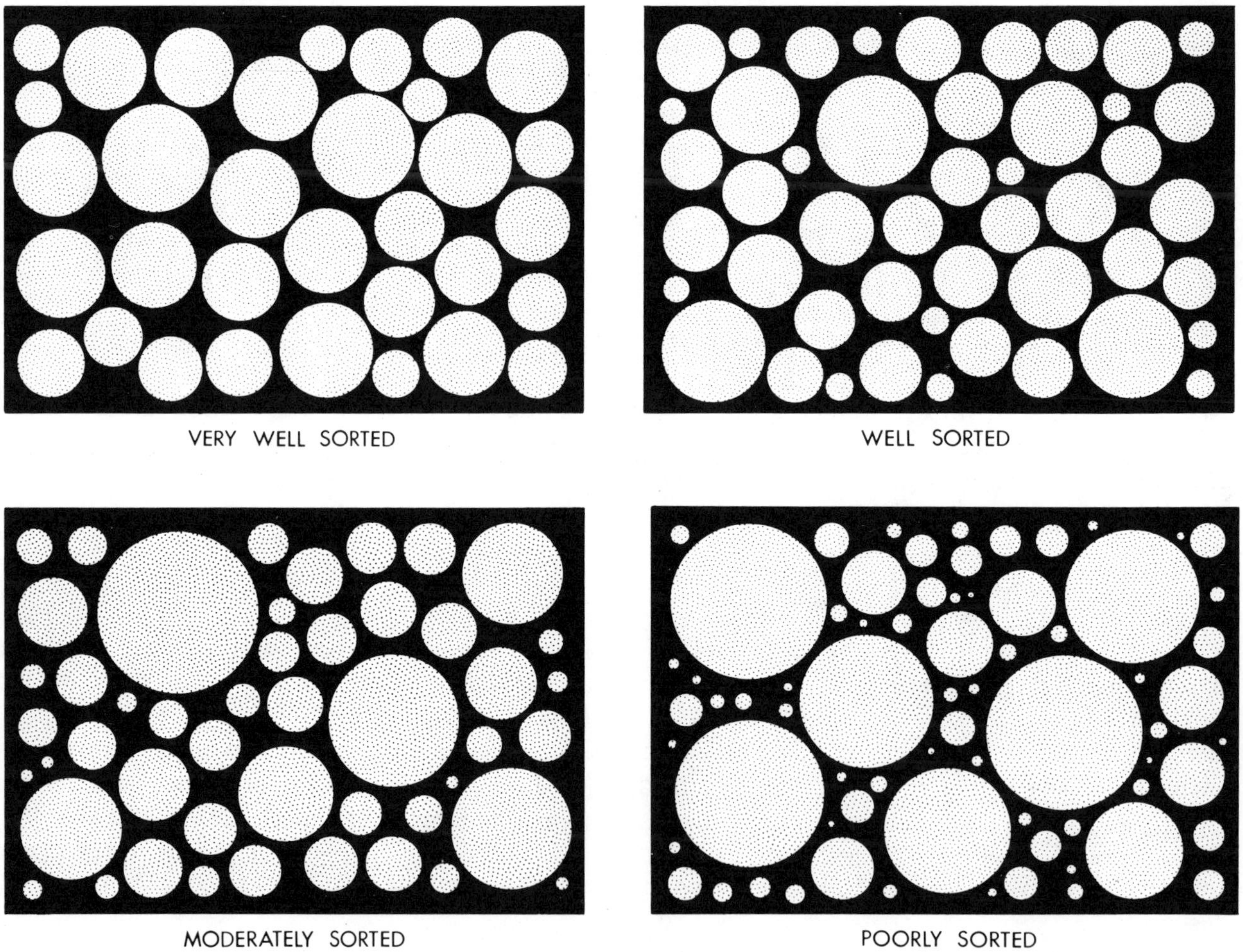

Figure 1.2. Grain sorting images.

smallest diameters in each of the four sorting images shown in Figure 1.2. What can cause size variation within a sediment? One possibility is that the sediment either has not been transported far or has not been extensively worked by water, giving it a mixture of grain sizes similar to that of the untransported, unworked soils weathering from the parent rock. Very poor sorting is found in alluvial fans and glacial tills, in the former because of lack of transport and in the latter because there has been no working by water. Another possibility is that a sudden loss of kinetic energy by the transport medium caused a wide range of grain sizes to be deposited simultaneously. This happens when a stream has a sharp change in gradient (as when flowing over an escarpment) or when it loses velocity in flowing around a meander loop and deposits poorly sorted sediment on the inside margin of the curve. Conversely, a relatively constant or high level of kinetic energy, as in waves breaking on a beach, produces a well-sorted sediment (see Figure 1.4).

Figure 1.3 shows the grain size distributions found in twelve major depositional environments. As an exercise, rank the environments in order of predominant grain size and in order of sorting. What physical processes cause the sorting, or lack of it, in each environment?

Grain size distributions are commonly represented by histograms (bar graphs) in which the height of each bar indicates the percentage of the total sediment (by weight) that each size interval accounts for. Figure 1.4 illustrates the sorting of four common sediment types.

Exercise

Sediment samples from different depositional environments will be provided in the laboratory.

1. Before beginning the analysis, write down the predicted roundness, grain size, and sorting for each sediment, based upon your knowledge of the sedimentary processes that characterize each environment.
2. Examine each sediment under a binocular microscope to test for roundness. Using Figure 1.1, determine the roundness of 10 randomly selected grains. What is the *dominant* roundness of each sample?
3. Use the sediment splitter to obtain about 40 grams of each sample. Separate the sediment into size fractions, using an appropriate collection of nested sieves. (Your instructor will explain the technical details of sieve analysis.) Although sieving does not yield precise results, it will give a rough approximation of the sediment size distribution. Weigh each size fraction and present the results as a histogram similar to those in Figure 1.4.
4. Compare your results with your predictions, and discuss possible reasons for any disagreement that you might have found.

Questions for Discussion

1. Does any correlation necessarily exist between the size, sorting, and roundness observed within a sediment? What could cause such a relationship?
2. What could cause grains of different size to have different degrees of roundness within the same sediment?
3. Is there any relationship between the age of a sediment particle (time since erosion) and its size? Could different particles within the same sediment be of different ages?
4. What could cause the presence of well-rounded grains in a poorly sorted sediment?
5. Can all sedimentary environments be differentiated by the size and shape of their grains? Which environments are distinctive? Which are not?

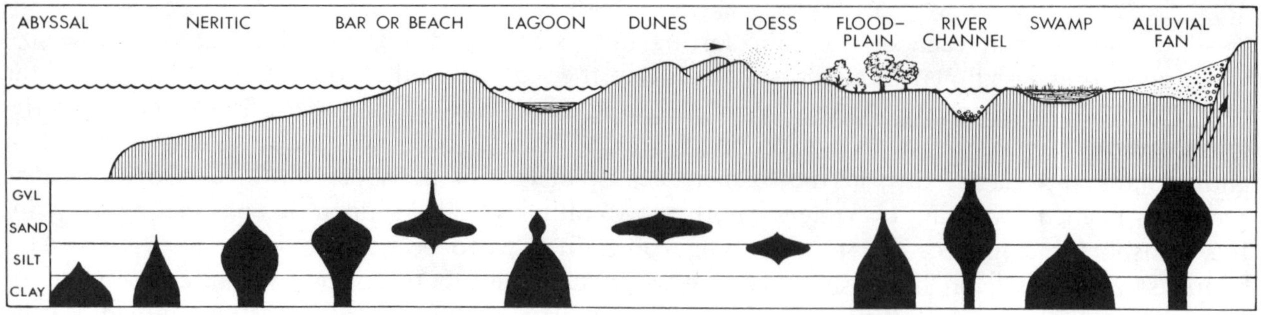

Figure 1.3. Grain size distributions in major depositional environments (very generalized).

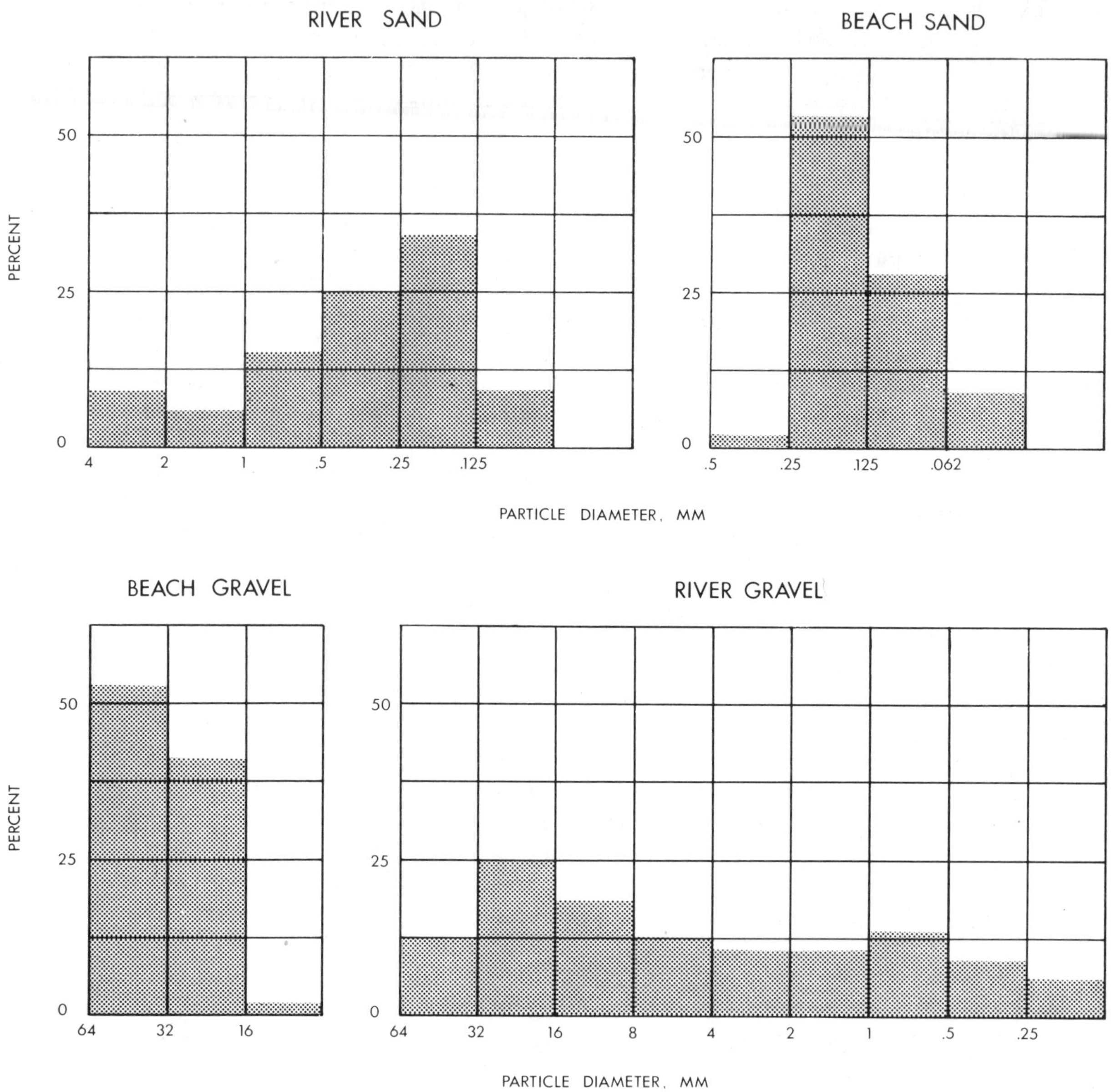

Figure 1.4. Grain size distributions of beach and river sediments.

Background Reading

Blatt, H., Middleton, G., and Murray, R. 1972. Origin of Sedimentary Rocks, pp. 32-78. Englewood Cliffs, N.J.: Prentice-Hall.

Folk, R. L. 1965. Petrology of Sedimentary Rocks, pp. 1-52. Austin, Tex.: Hemphill's.

Krumbein, W. C., and Sloss, L. L. 1963. Stratigraphy and Sedimentation, 2nd ed., pp. 96-114. San Francisco: W. H. Freeman.

Chapter Two

Sedimentary Petrology
Origin of Detrital (Clastic) Rocks

Objectives

1. To recognize the grain and aggregate properties of the major types of detrital sedimentary rocks.
2. To interpret the origins of these rocks in terms of source materials, transport, and depositional environment.

Important Terms

Chemical—Derived by chemical or biochemical processes in the place of deposition and not transported.

Detrital—Derived by the weathering and transportation of a preexisting rock. The term *clastic* is sometimes substituted.

Erosion—The combined process of the weathering of parent rock material and transport of the detritus away from the place of origin.

Tectonics—Crustal readjustment causing rock deformation through faulting or folding. Regions of tectonic activity are the main source areas of detrital sediments.

Introduction

The formation of detrital rocks requires that streams and rivers transport large amounts of eroded detritus to the depositional site. Detrital rocks indicate the existence of a contemporary actively eroding source area, most likely the result of some kind of tectonic activity. The origin of chemical rocks requires that the supply of detritus be small, through either the general absence of eroding source areas or the local exclusion of detritus from the depositional area. All three major rock classes may have formed in the same area in deltas, marine shorelines, and in lakes. (See Table 2.1.)

TABLE 2.1. Major Categories of Sedimentary Rocks

Class	*Common Rock Types*	*Source of Grains*
Detrital	Mudstone or shale Siltstone Sandstone or graywacke Conglomerate or breccia	Erosion of preexisting igneous, metamorphic, and/or sedimentary rocks
Chemical	Limestone Dolomite (dolostone) Evaporites Coal	Plant and animal skeletons or tissues or direct precipitation from water
Mixed	Shaley limestone Organic shale Sandy dolomite Limy siltstone	Material from both sources above

Detrital Minerals

The types of mineral grains in a detrital sediment depend upon the composition of the source rocks as well as the size of the particles and the length of time that they have been subject to further decomposition. The stability of a mineral species is roughly the inverse of its original temperature of crystallization. Stability generally fits into the pattern shown in Table 2.2.

TABLE 2.2. Relative Stability of Ferromagnesian Silicates and Aluminosilicates

Stability	*Ferromagnesian Silicates*	*Aluminosilicates*
Lowest	Olivine	Calcium plagioclase
↓	Pyroxene	Sodium plagioclase
	Amphibole	Potassium feldspar
	Biotite	Muscovite
	Limonite/Hematite	Quartz
Highest	Clay minerals	Clay minerals

The most stable minerals in sediments are clays, quartz, iron oxides, and micas. Feldspars and ferromagnesian silicates are usually rare, except in sediments that have neither been transported far nor extensively worked by water, as in alluvial fan deposits. The most common detrital rocks are sandstones and mudstones (or shales), usually stained with iron oxides.

The mineralogy of a detrital rock may also be a key to its depositional environment. A stream-deposited sand may contain feldspars, micas, and rock fragments, whereas beach sands from the same source may consist almost entirely of quartz. How can these two different environments affect the mineralogy of their sediments?

Grain Properties

Grain roundness has been treated in Chapter 1 (Figure 1.1.). Grain size reflects the velocity of the transporting wind or water at the time of deposition. The water on a dirt road after a rainstorm moves clay and silt; the water on a beach moves sand and pebbles. Thus sandstones and conglomerates accumulated in streams and on beaches, and shales and mudstones were deposited in swamps, quiet bays, floodplains, or lagoons (Figure 1.3). Mixtures of size classes or intermediate

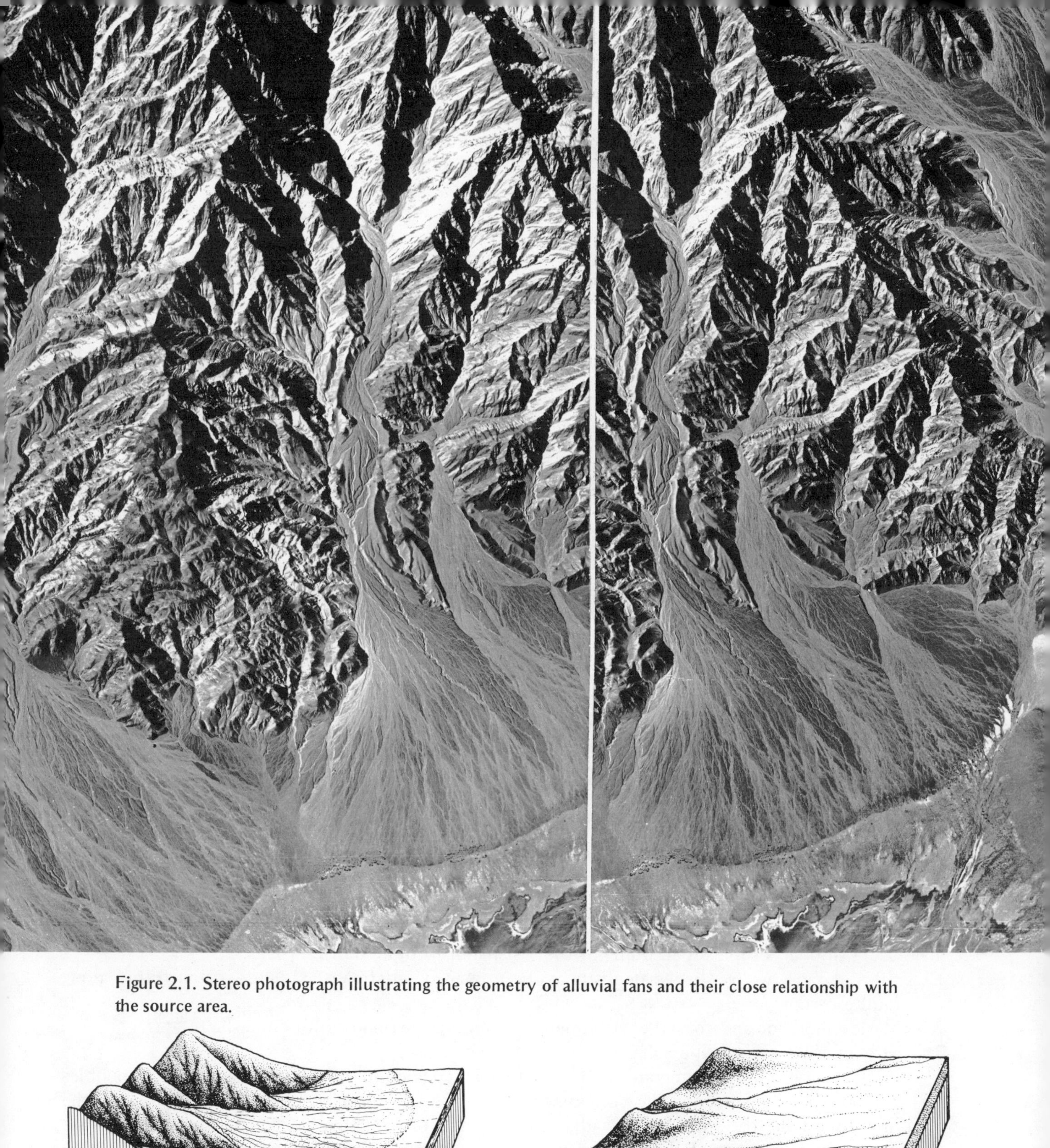

Figure 2.1. Stereo photograph illustrating the geometry of alluvial fans and their close relationship with the source area.

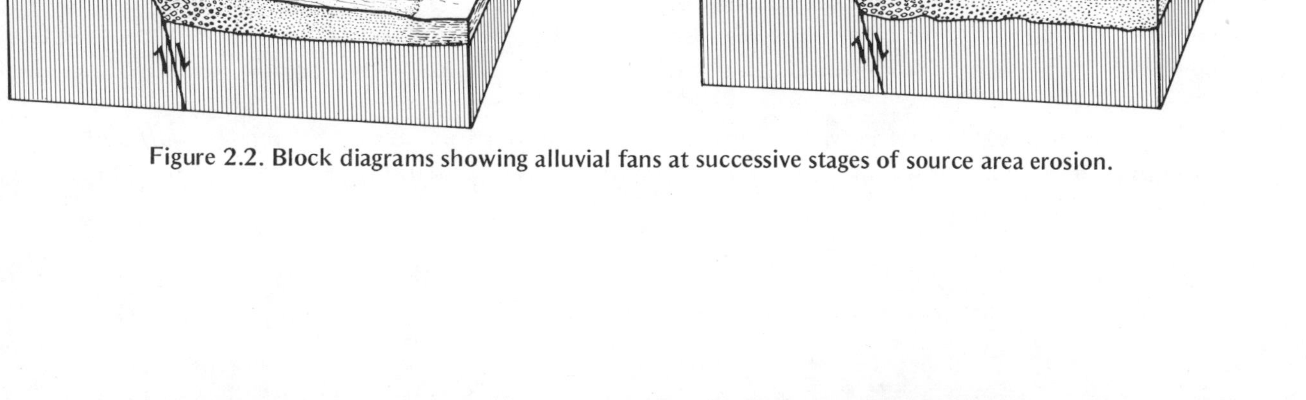

Figure 2.2. Block diagrams showing alluvial fans at successive stages of source area erosion.

grain sizes reflect intermediate conditions or the gradual encroachment of one environment on another. Detrital rocks and sediments are classified according to their most common grain size (Table 2.3).

TABLE 2.3. Grain Size of Sediments in Detrital Sedimentary Rocks

Grain Diameter (Millimeters)	*Sediment*	*Rock Name*
2 to 64	Gravel	Conglomerate
0.06 to 2	Sand	Sandstone
0.004 to 0.06	Silt	Siltstone
Less than 0.004	Clay	Claystone or shale

Aggregate Properties

Sorting has been discussed in Chapter 1 (Figures 1.2, 1.3, and 1.4). Because sorting reflects water velocity, which is normally fairly constant over a period of time, many sediments are moderately well sorted. Poorly sorted sediments reflect variations in water velocity, such as occur during floods or storms or in water flowing over an escarpment.

Much of the volume of some rocks comes from the organic tissues and skeletons of plants and animals. Fine-grained sediments contain considerable organic matter, which can give the rock a black color. Such material generally consists of free carbon, hydrocarbons, and other organic compounds. It is very common in shales. Why is this material much less abundant in sandstones?

Commonly, organic material can be recognized as parts of once-living plants and animals, called *fossils.* Fossils constitute a considerable percentage of the volume of detrital rocks (up to an arbitrary level of 50 percent) and the bulk of most carbonate rocks and coal. Knowledge of the life habits of similar living organisms may permit interpretation of the conditions in which the fossiliferous rocks were formed.

An extended discussion of the aggregate properties of sedimentary rocks is found in Chapters 8, 9, and 10, which deal with sedimentary structures.

Depositional Environments

The *depositional environment* is the area over which a particular rock type accumulates. Three major depositional complexes account for the vast majority of all detrital rocks.

1. Alluvial fans (wedge-shaped deposits at sharp changes in stream gradients, often found along fault escarpments).
2. Marine shoreline and coastal environments, including the open continental shelf.
3. Stream, river, and marine delta systems, including lakes.

Coarse, unsorted conglomerates and mixed rock types containing much clay, silt, and pebbles are common in alluvial fans. The conglomerate pebbles and rock fragments directly reflect the materials found on the immediately adjacent eroding upland (Figure 2.1). Coarse materials are deposited nearer the head of the fan and finer sediment more towards the base (Figure 2.2). Near the base, sorting improves so that the sediment more nearly resembles typical stream deposits.

Padre Island, along the Texas coast (Figure 2.3), is a classic example of coastal sedimentation, where the buildup of sand is influenced mostly by longshore currents. The barrier island consists of beach and dune sand, while mud and silt accumulate in the bay, lagoon, and marine shelf environments.

The birdfoot delta of the Mississippi River is a standard example of river (*fluvial)* and deltaic sedimentation (Figures 2.4 and 2.5). In deltas many types of detrital rocks accumulate in proximity

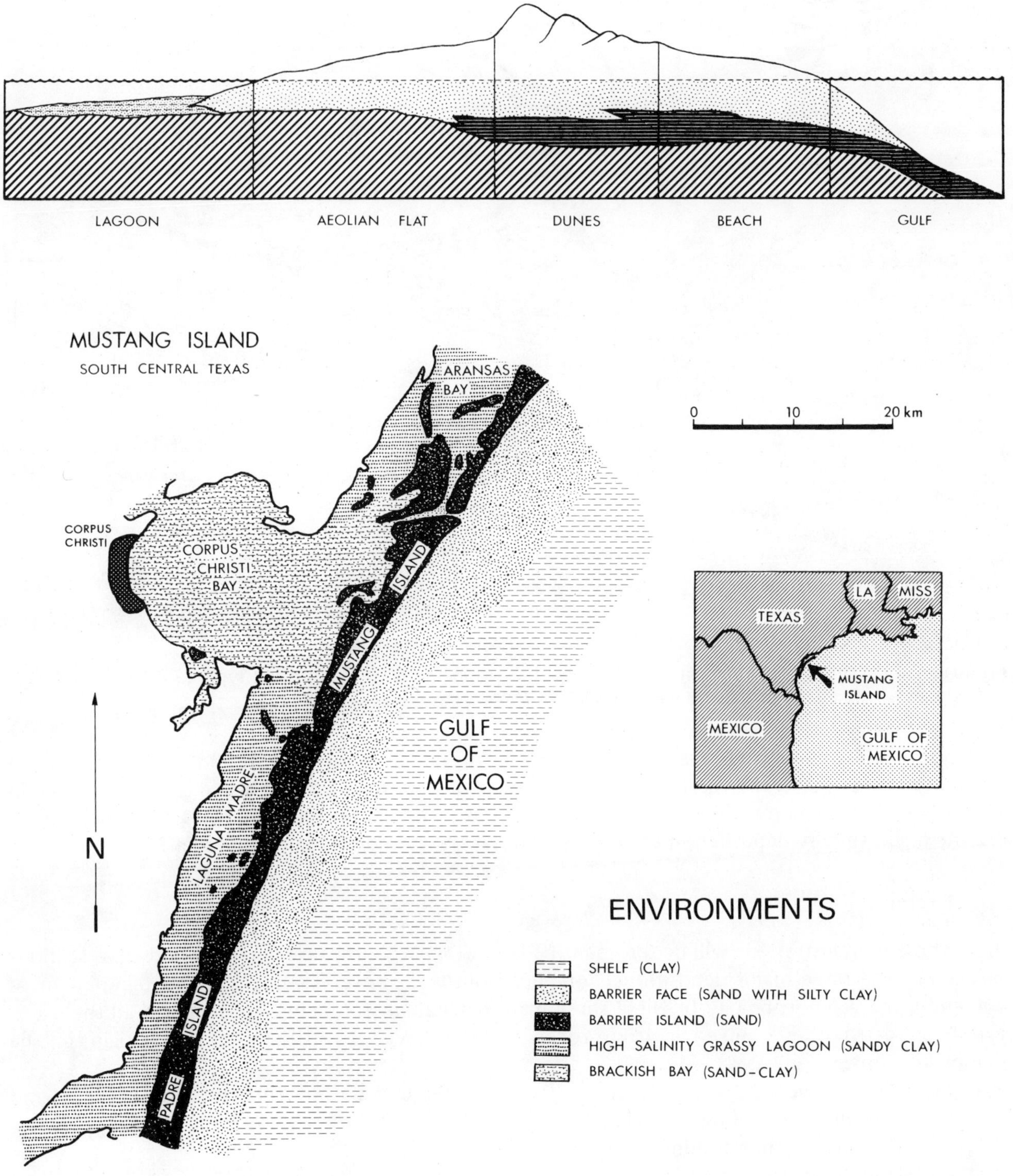

Figure 2.3. Depositional environments associated with a barrier island shoreline.

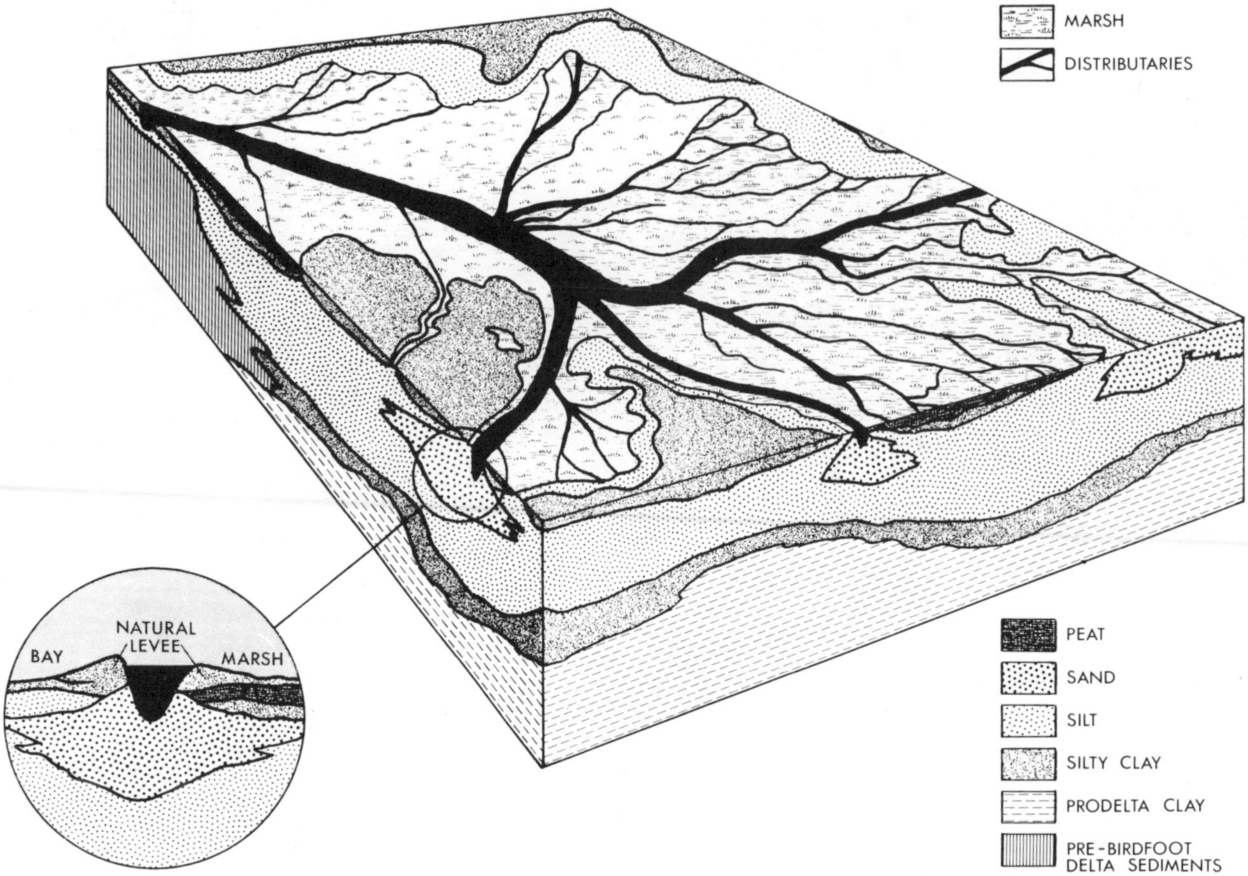

Figure 2.4. Depositional environments associated with a marine delta of the birdfoot type; to the lower left is a vertical cross section through a distributary mouth bar.

and often intergrade. Sands are deposited on the inside curves of river meander loops, along the channel margins (natural levees), and at the mouths of distributary channels. Silts and clays rich in organic materials accumulate on the flood plains and in the swamps and marshes. Clays containing marine organisms are deposited in the interdistributary bays and on the marine shelf.

Exercise

Three demonstrations will be set up in the laboratory to illustrate map views and cross sections of each of the three major depositional complexes outlined above. Also, selected topographic maps will show the geomorphology of these three main detrital environments. Examples of all the major types of detrital and mixed detrital rocks will be keyed to their places of origin in the demonstrations.

1. You will receive a set of unidentified detrital rocks. Using a binocular microscope or a hand lens, determine the following for each unknown:
   a. Dominant grain roundness.
   b. Dominant grain size.
   c. Sorting.
   d. Organic content.

Figure 2.5. The birdfoot delta of the Mississippi River.

2. Determine to which major environmental complex each specimen *most likely* belongs. Make sketches of the map views and cross sections in the demonstrations and indicate the probable position(s) of origin of each unknown sedimentary rock.

Questions for Discussion

1. Which detrital rock types are highly diagnostic of particular depositional environments and which could have been deposited in any of several environmental settings?
2. Which general types of fossils would be particularly useful in determining the depositional environment of some of the less diagnostic rock types?
3. If the rocks of the source area above an alluvial fan had a horizontal stratification of mineral assemblages, in what order would this mineralogical sequence be observed in the fan sediments?
4. Explain why brackish water and water of very high salinity are found only in certain depositional environments.
5. What effects will large storms or hurricanes have on the depositional patterns discussed in this chapter? How often should such disruptions be expected in the sequence of sediments representing the last million years in the area of the Mississippi delta?

Background Reading

Blatt, H., Middleton, G., and Murray, R. 1972. Origin of Sedimentary Rocks, pp. 185-214. Englewood Cliffs, N.J.: Prentice-Hall.

Bull, W. B. 1972. "Recognition of Alluvial Fan Deposits in the Stratigraphic Record." In J. K. Rigby and W. K. Hamblin, eds., Recognition of Ancient Sedimentary Environments, pp. 63-83. Society of Economic Paleontologists and Mineralogists, Special Publication 16.

Krauskopf, K. G. 1967. Introduction to Geochemistry. New York: McGraw-Hill.

Selley, R. C. 1970. Ancient Sedimentary Environments, pp. 22-51, 74-116. Ithaca, N.Y.: Cornell University Press.

Chapter Three

Paleontology
Invertebrate Fossils

Objectives

1. To learn to identify the most common types of invertebrate fossils.
2. To become acquainted with the overall comparative anatomy of the animal kingdom.
3. To understand the basic life habits and living conditions of invertebrates commonly found as fossils.

Important Terms

Fossils—The preserved remains, usually skeletal fragments, of once-living organisms found buried in ancient sediments or as the constituent grains of sedimentary rocks.

Habitat—The place (either surface or volume) that an organism occupies during life.

Life habits—The ways in which an organism goes about its daily activities, primarily those of locomotion and food gathering.

Introduction

The places where sediment accumulates support particular and often unique types of plants, animals, and microorganisms. Those that have hard shells, skeletons, or woody fiber may be buried in the sediment and preserved as fossils. The assemblages of fossils found in particular beds of rock have proven to be of considerable usefulness in deciphering the characteristics of the ancient environment, including features that may not be recorded in any other way. Unusual mixtures of fossils in some rocks may indicate unusual conditions of sedimentation, such as the way a storm or hurricane can mix together animals from both deep and shallow water. Knowledge of the ancient environment in which an animal lived might enable a paleontologist to interpret puzzling aspects of its shape or form as environmental adaptations and thus understand better how it evolved. Aspects of morphology that are known to be environmental adaptations may likewise tell us something about an ancient habitat. The assemblage of fossils found in a rock includes not only the animals that lived on the seafloor or lake bottom, but also those that lived buried in the sediment, that floated or swam in the water, that inhabited nearby land areas, and that flew or were suspended in the overlying air.

Overall Classification of Organisms

Biologists have created a filing system for talking about organisms and understanding their evolutionary relationships. Although some of the pigeonholes are a bit arbitrary, most of them contain groups of organisms that are united by common aspects of form, function, and complexity. In addition, the files are arranged in a hierarchy, so that each large category is subdivided into nestings of smaller categories. The largest groups are the five *kingdoms*, illustrated in Figure 3.1 against a geologic time scale. The kingdoms are broken down into *phyla* (singular, phylum), the phyla into *classes*, the classes into *orders*, the orders into *families*, the families into *genera* (singular, genus), and the genera into the smallest category, the *species* (singular and plural, *species)*. In addition, many intermediate groupings are also used, such as superphylum, subclass, superfamily, and so forth. Some biologists claim that only the species has biological reality and that the other categories are more or less artificial, established upon arbitrary criteria.

The five kingdoms comprise the *monerans, protists, fungi, plants*, and *animals*. Monerans are very small single cells without nuclei (procaryotes). They include the bacteria and the blue-green algae. Monerans are geologically important because they formed laminated structures called *stromatolites*, which are very common in carbonate and siliceous rocks. (Stromatolites are discussed further in Chapter 10.) The protists are larger single cells with a nucleus, a feature they have in common with their multicellular relatives, the fungi, plants, and animals (all eucaryotes). The protists include many types of algae, some of which are important in forming limestones (see Chapter 7), and protozoans, some of which developed shells and are important as microfossils. The plants and fungi are multicelled organisms with very stiff, rigid cell walls, whereas the animals are multicelled organisms with relatively soft, pliable cells.

The animal kingdom is divided into about 25 phyla, of which only 9 or so have an adequate fossil record. The identification key will help you to identify 7 phyla and 10 classes of macroscopic invertebrates that are commonly found as fossils. A course in invertebrate paleontology will introduce you to many more. Perhaps the most important development among animals was that of an internal body cavity, the *coelom*, a fluid-filled chamber that holds most of the important organs. The lower invertebrates (sponges, jellyfish, and corals) are very simple animals without a coelom or well-developed organ systems. The coelomates are divided into two groups, possibly artificial, based on early patterns of growth of the embryo. In the *protostomes* (molluscs, annelids, and arthropods) the embryonic mouth becomes the mouth of the mature animal. In the *deuterostomes* (lophophorates, echinoderms, and chordates) the embryonic mouth becomes the anus, and a second mouth appears that becomes the mouth of the mature animal. The animals in the identification key are classified in the following way.

- Phylum PORIFERA
  - Class SCLEROSPONGIAE
    - Order STROMATOPORIDA (stromatoporoids, extinct animals probably related to living sclerosponges; Figure 3.2)
- Phylum CNIDARIA
  - Class ANTHOZOA (corals, both solitary and colonial; Figure 3.2)
    - Order TABULATA (tabulate corals, some of which—the favositids—may not be corals or cnidarians at all, but could be related to living sclerosponges; Figure 3.2)
- Phylum MOLLUSCA
  - Class PELECYPODA (clams, oysters, scallops; Figure 3.3)
  - Class GASTROPODA (snails, limpets; Figure 3.3)
  - Class CEPHALOPODA

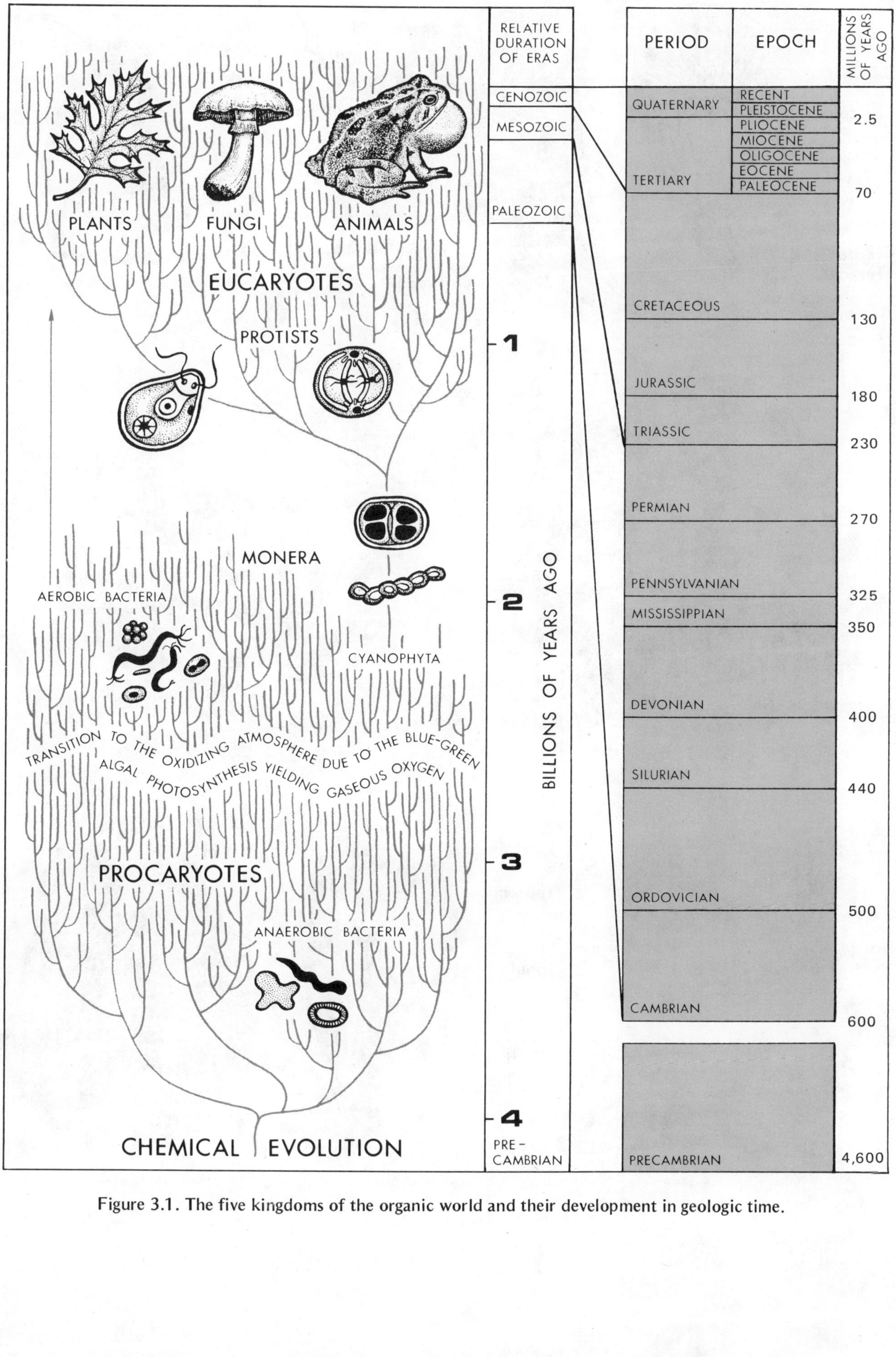

Figure 3.1. The five kingdoms of the organic world and their development in geologic time.

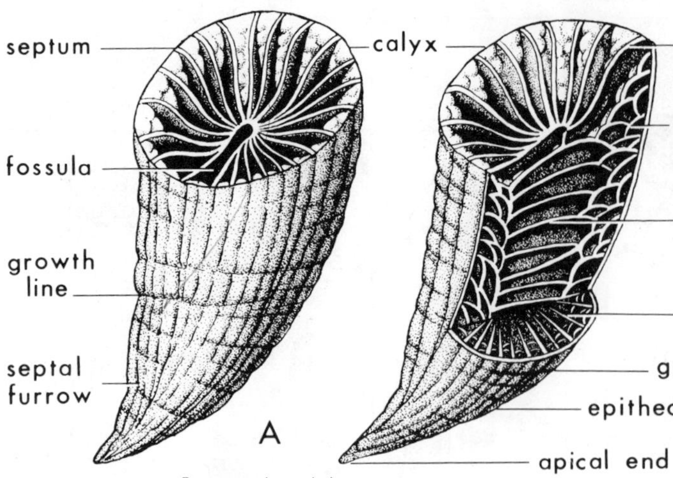

Rugose coral morphology

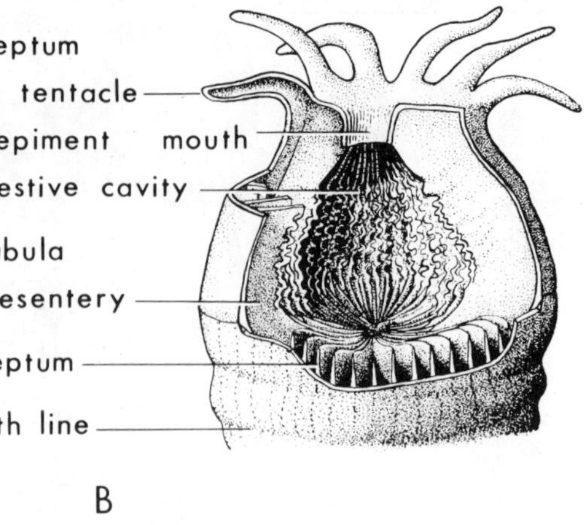

Living rugose coral, reconstructed

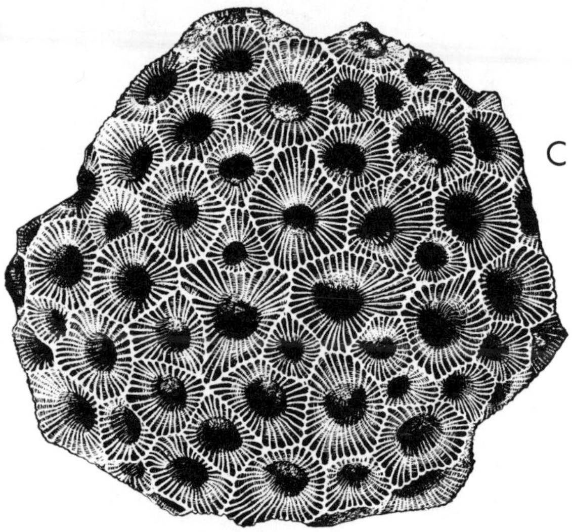

Hexagonaria, a Devonian colonial rugose coral

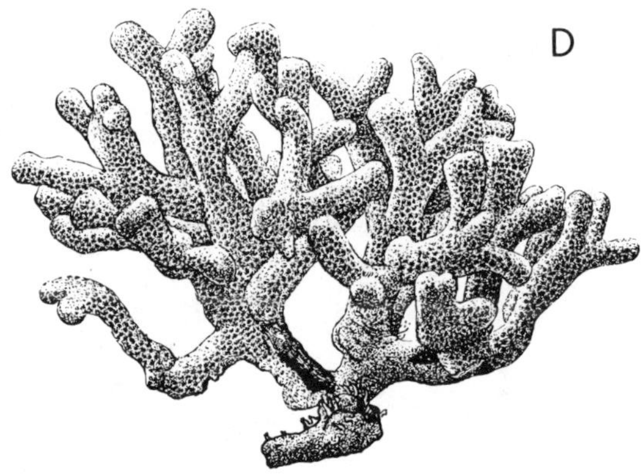

Porites, a Recent colonial coral (scleractinian) (X.6)

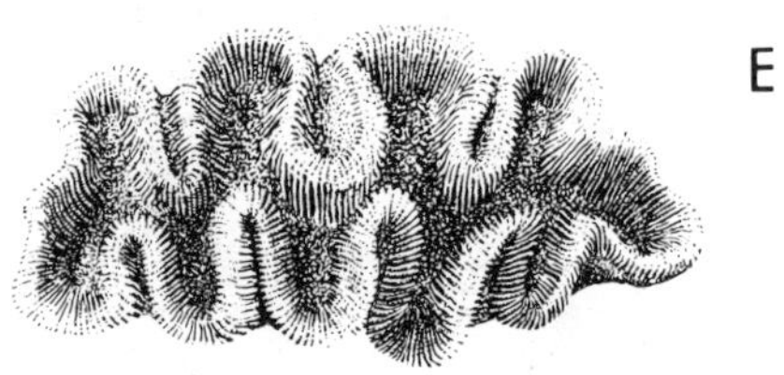

Manicina, a Recent "solitary" coral (scleractinian) (X.6)

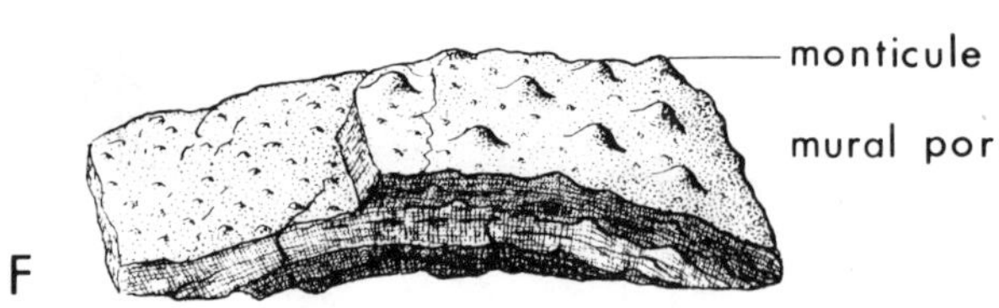

Stromatoporoid, Devonian

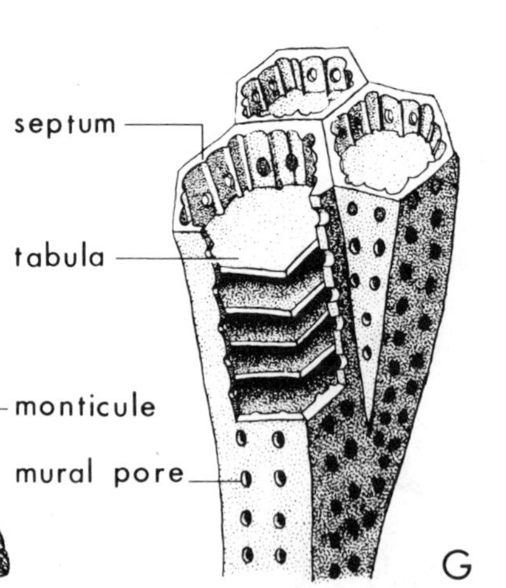

Tabulate coral morphology (enlarged)

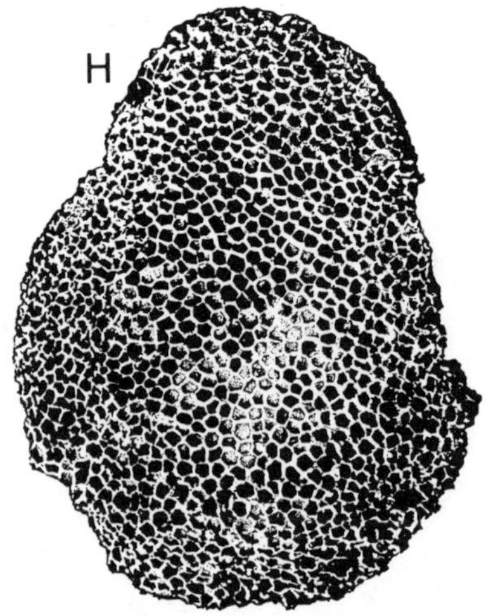

Favosites, a Devonian tabulate coral (X.6)

Figure 3.2. Lower invertebrates.

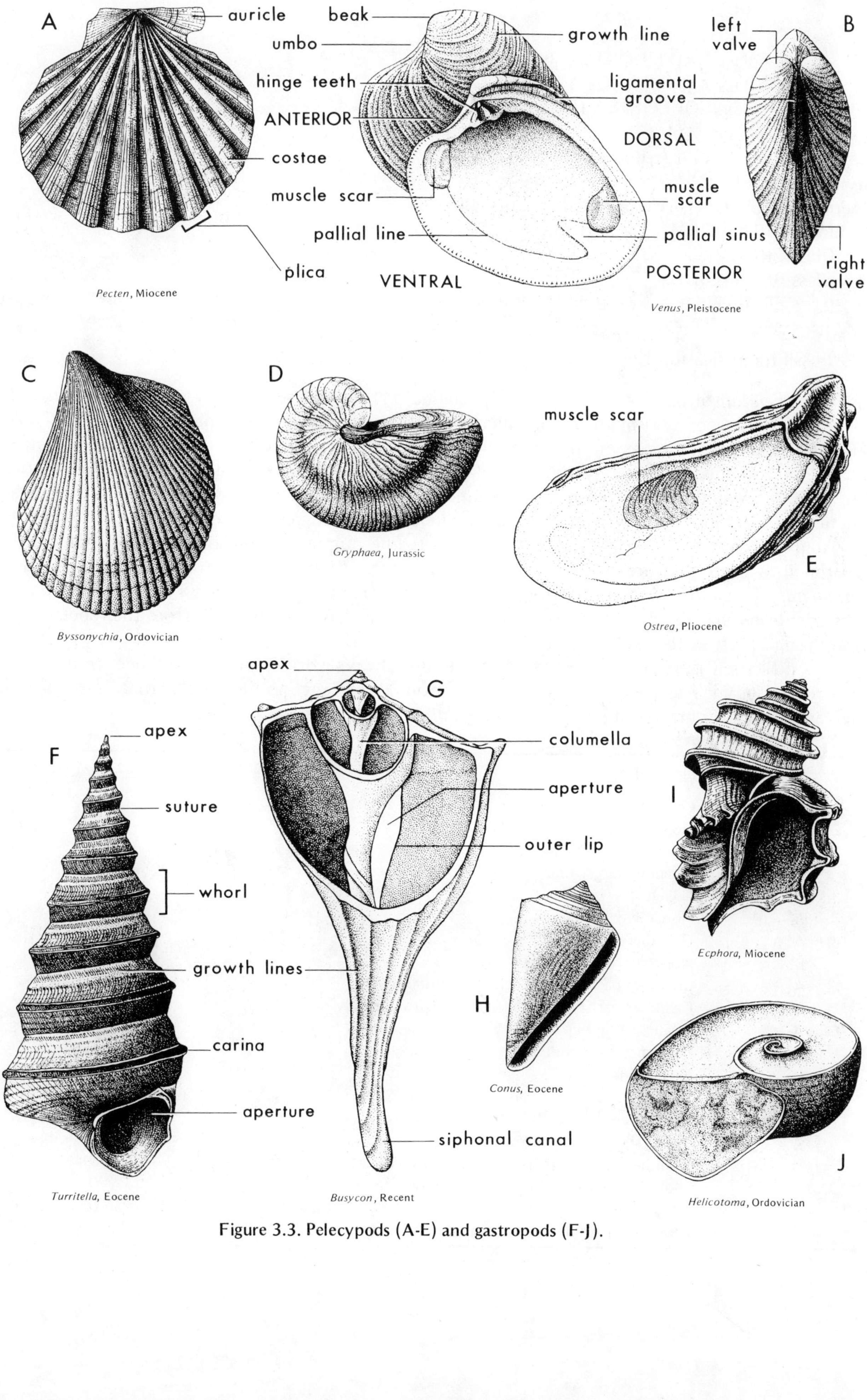

Figure 3.3. Pelecypods (A-E) and gastropods (F-J).

Subclass NAUTILOIDEA (Figure 3.4)
Subclass AMMONOIDEA (Figure 3.4)
Subclass COLEOIDEA (Figure 3.4)
Phylum ARTHROPODA
Class TRILOBITA (trilobites; Figure 3.5)
Phylum ECTOPROCTA (bryozoans; Figure 3.6)
Phylum BRACHIOPODA (lampshells; Figure 3.6)
Phylum ECHINODERMATA
Class CRINOIDEA (sea lilies; Figure 3.7)
Class BLASTOIDEA (Figure 3.7)
Class ECHINOIDEA (sea urchins, sand dollars; Figure 3.5)

Fossil Identification Key

The *skeleton* of an animal consists of special hard structures secreted by the animal, usually of calcium carbonate or phosphate or organic tissues that have been impregnated with minerals. An *exoskeleton* surrounds most of the other tissues and organs, whereas an *endoskeleton* is totally encased by the body. Most invertebrates have exoskeletons; an exception are the belemnoids (Figure 3.4). A single exoskeleton having numerous tubes or openings that each house a separate functioning organism is *colonial.* A tubular exoskeleton that curves around an axis of rotation (like a twisted garden hose) is *coiled. Symmetry* refers to the number of imaginary planes that cut the skeleton into parts that are mirror images of each other. *Bilateral* symmetry has one such plane; *pentameral,* five; and *radial,* an infinite number. *Chambers* are formed when a tubular skeleton has cross partitions that divide it into a series of hollow sections. *Segments* result from the repetition of discrete solid parts of the skeleton.

Identification keys are very artificial and do not always work for every specimen. In using this key it is a good idea to check your results by continual reference to Figures 3.2 to 3.7. Identifying fragments or poorly preserved material is always difficult.

Key Category		*Description*	*Organic Group*
1.	A.	Colonial	Go to 2
	B.	Noncolonial	Go to 3
2.	A.	Openings smaller than 1 mm	Go to 13
	B.	Openings larger than 1 mm	Go to 14
3.	A.	Coiled	Go to 4
	B.	Noncoiled	Go to 6
4.	A.	Chambered	Go to 5
	B.	Nonchambered, hollow	Gastropods
5.	A.	Chamber partitions straight or gently curved	Nautiloids
	B.	Chamber partitions highly curved with "crinkly" edges	Ammonoids
6.	A.	Bilateral symmetry	Go to 7
	B.	Pentameral symmetry	Go to 9
	C.	Radial symmetry	Go to 11
7.	A.	Two paired shells (valves)	Go to 8
	B.	Segmented, having three lobes	Trilobites
8.	A.	Paired shells exact mirror images	Pelecypods
	B.	Shell halves not mirror images; each valve has bilateral symmetry	Brachiopods
9.	A.	Globular or vaselike, covered by polygonal plates; stem remnants or attachment marks at base	Go to 10
	B.	Globular or disklike, not connected to a stem	Echinoids

Key Category		Description	Organic Group
10.	A.	Five radiating elaborate grooves	Blastoids
	B.	No elaborate grooves	Crinoids
11.	A.	Conical or tapering shell	Go to 12
	B.	Cylindrical, composed of disklike segments	Crinoid stems
12.	A.	Horizontal partitions forming chambers	Go to 5
	B.	Vertical partitions radially projecting inward	Solitary corals
	C.	Shell solid, cigarlike, having radial orientation of fibrous crystals	Belemnites
13.	A.	Massive, with thin laminations, each thin layer having numerous boxlike chambers	Stromatoporoids
	B.	Not thinly laminated	Bryozoans
14.	A.	Vertical partitions radially projecting inward within each opening	Colonial corals
	B.	Horizontal partitions forming chambers	Tabulate corals

Life Habits and Habitats

The major environments in which animals live are shown in Figure 3.8. Animals and plants that live in air, on land, or in fresh water are shown in Figures 3.10 to 3.13. Organisms that float freely in the water (usually near the water's interface with the air) are called *plankton* (Figure 3.14). Animals that swim (control their own movement in the water) are called *nekton* (Figure 3.15). Animals and plants intimately associated with the substratum, either unconsolidated sediment or a hard rocky seafloor, are the *benthos.* Those that are attached or do not move about freely are *sessile,* whereas those that crawl about over the surface are *vagrant* (Figure 3.16). Benthic animals living above the sediment-water interface are the *epifauna* (Figure 3.17); those living within the sediment (burrowers) are the *infauna* (Figure 3.18). The interrelationships of these types of life habits (Figure 3.9), when considered along with the physical processes and characteristics of the habitat, constitute an *ecosystem.* An ecosystem is the collection of pathways along which food and energy are channeled by a group of organisms living together.

An important aspect of an ecosystem is the way that each species obtains food. Its feeding mechanisms and type of food determine an animal's role (*niche)* in its community. Plants are *autotrophs,* manufacturing their own food through photosynthesis, but all other organisms are consumers of food. Animals that depend upon plankton and suspended food are *filterers.* Organisms that mine organic debris that has accumulated within the sediment are *deposit feeders.* Animals that feed on plants are *herbivores,* and those that eat other animals are *carnivores.* Feeders that search for dead organic matter lying on top of the sediment are *scavengers.* Table 3.1 lists the most common feeding types and life habits of the marine organisms you have identified using the key.

The relative proportions of different feeding types is an important characteristic of marine ecosystems. Communities dominated by filterers must have had a stable and reliable supply of suspended plankton for food. The sediment, on the other hand, provides a food reservoir for deposit feeders that does not depend upon the stability of suspended food in the water. Ecosystems subject to fluctuating resources, such as the supply of dissolved nutrients from a river mouth, may have a dominance of deposit feeders. Offshore ecosystems on a marine shelf with stable supplies of suspended food may have a dominance of filterers. The ecosystems with the most stable food supply will even have a variety of filterers arranged at different heights above the substrate, like the balconies on an apartment house, thereby exploiting the suspended food supply to the maximum. The overall stability of an ecosystem is reflected by the degree to which available resources are partitioned by the inhabitants. This partitioning of available resources is called *niche diversity.*

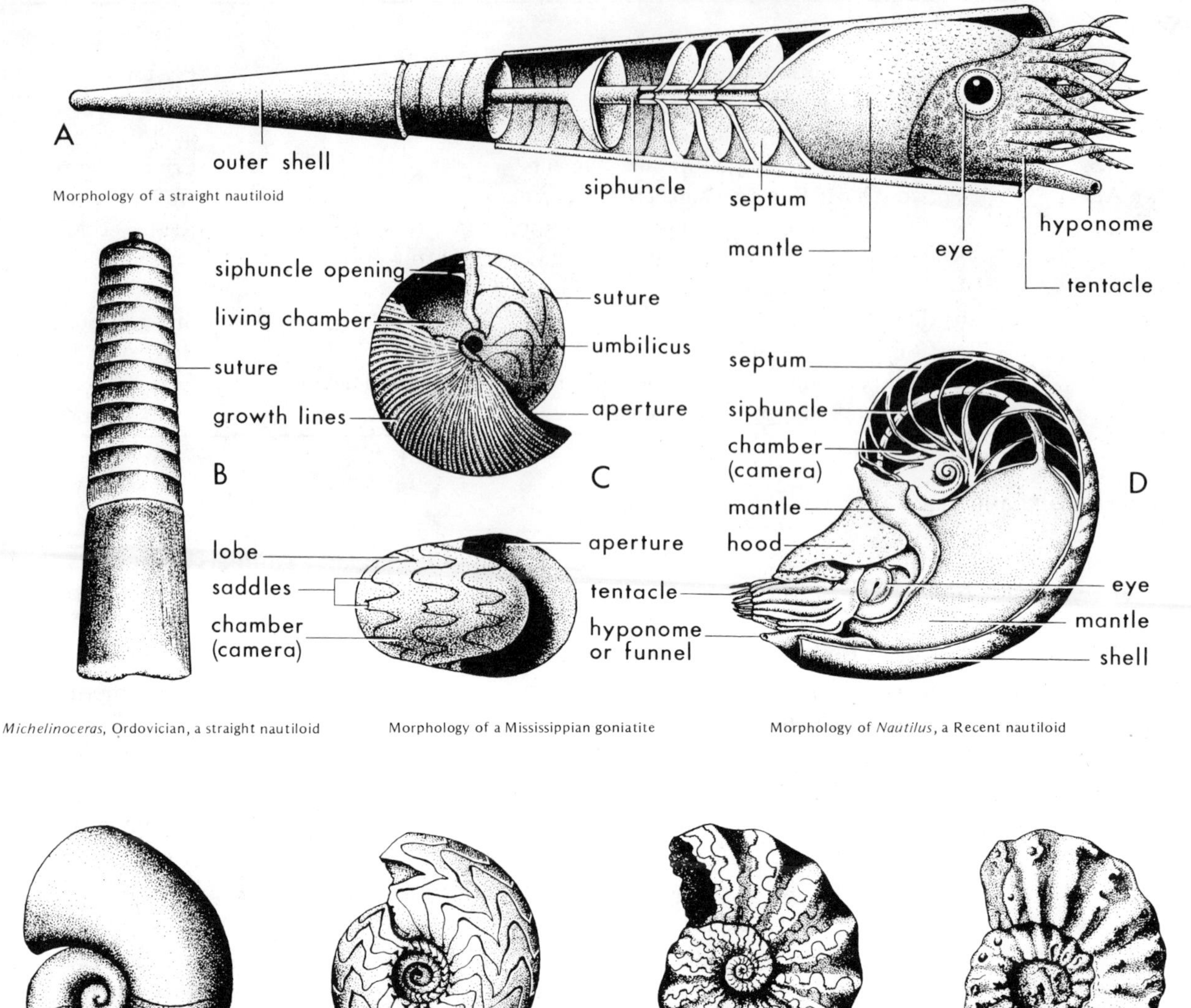

Figure 3.4. Cephalopods.

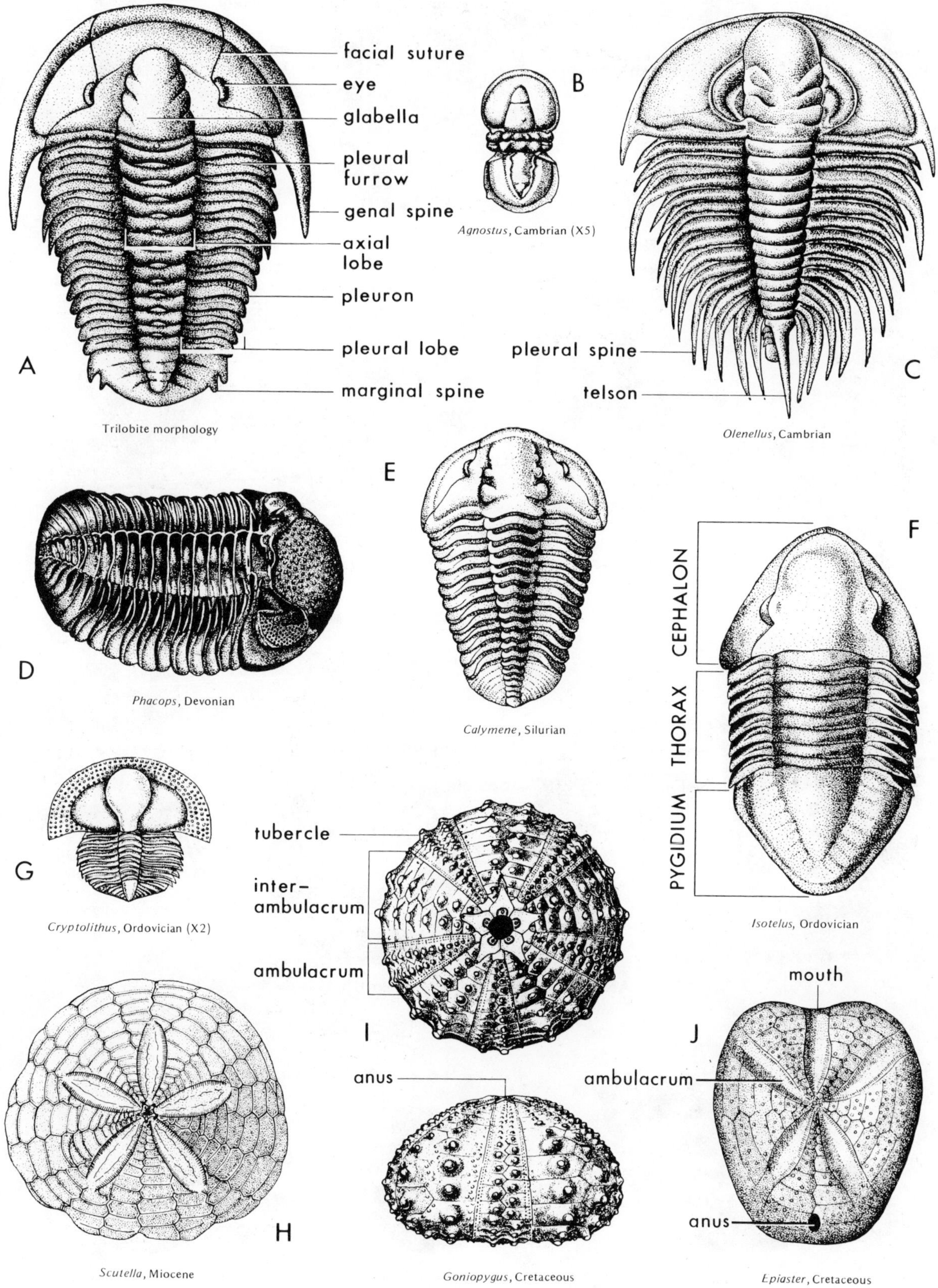

Figure 3.5. Trilobites (A-G) and echinoids (H-J).

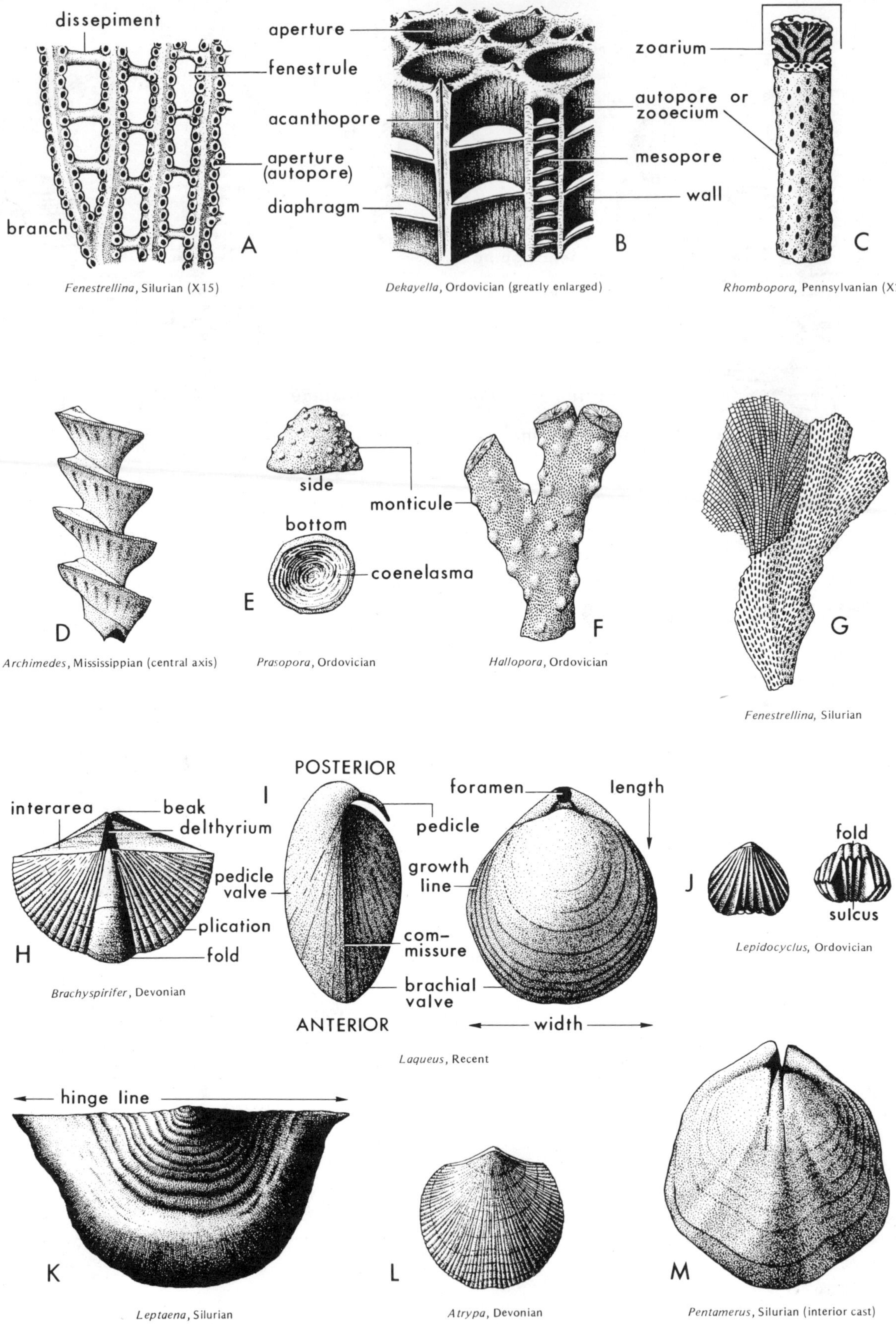

Fenestrellina, Silurian (X15)

Dekayella, Ordovician (greatly enlarged)

Rhombopora, Pennsylvanian (X2)

Archimedes, Mississippian (central axis)

Prasopora, Ordovician

Hallopora, Ordovician

Fenestrellina, Silurian

Brachyspirifer, Devonian

Laqueus, Recent

Lepidocyclus, Ordovician

Leptaena, Silurian

Atrypa, Devonian

Pentamerus, Silurian (interior cast)

Figure 3.6. Bryozoans (A-G) and brachiopods (H-M).

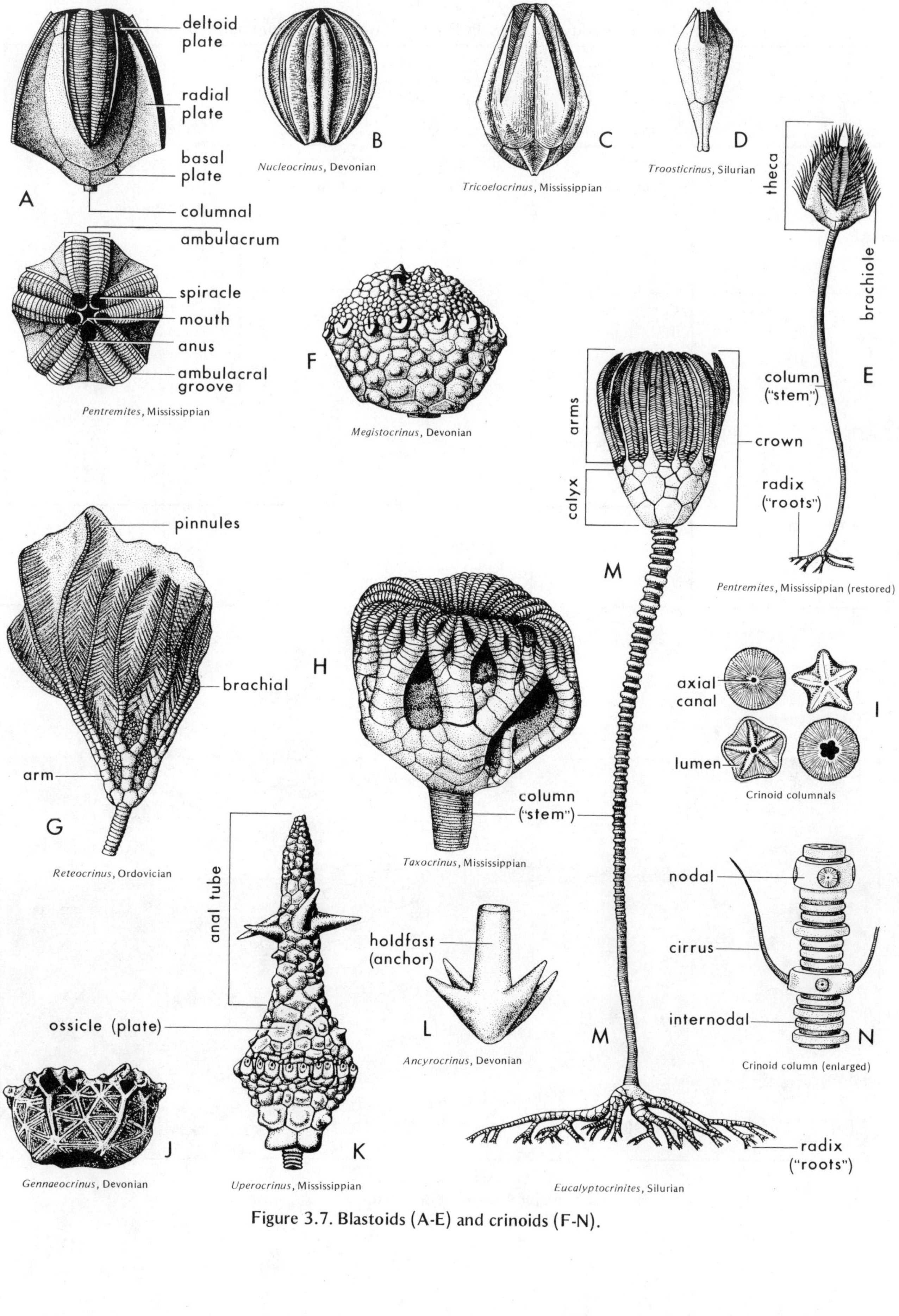

Figure 3.7. Blastoids (A-E) and crinoids (F-N).

TABLE 3.1. Life Habits and Feeding Types of Invertebrates Common as Fossils

Life Habits \ *Feeding Types*	*Filterers*	*Deposit Feeders*	*Herbivores*	*Carnivores/ Scavengers*
Vagrant epifauna	Gastropods Pelecypods	Trilobites Echinoids	Gastropods Echinoids	Gastropods Nautiloids Ammonoids Trilobites Echinoids Crinoids
Sessile epifauna	Corals Tabulates Stromatoporoids Bryozoans Brachiopods Pelecypods Gastropods Crinoids Blastoids			Corals
Infauna	Pelecypods Brachiopods	Echinoids Trilobites Pelecypods Gastropods		Gastropods Trilobites Echinoids

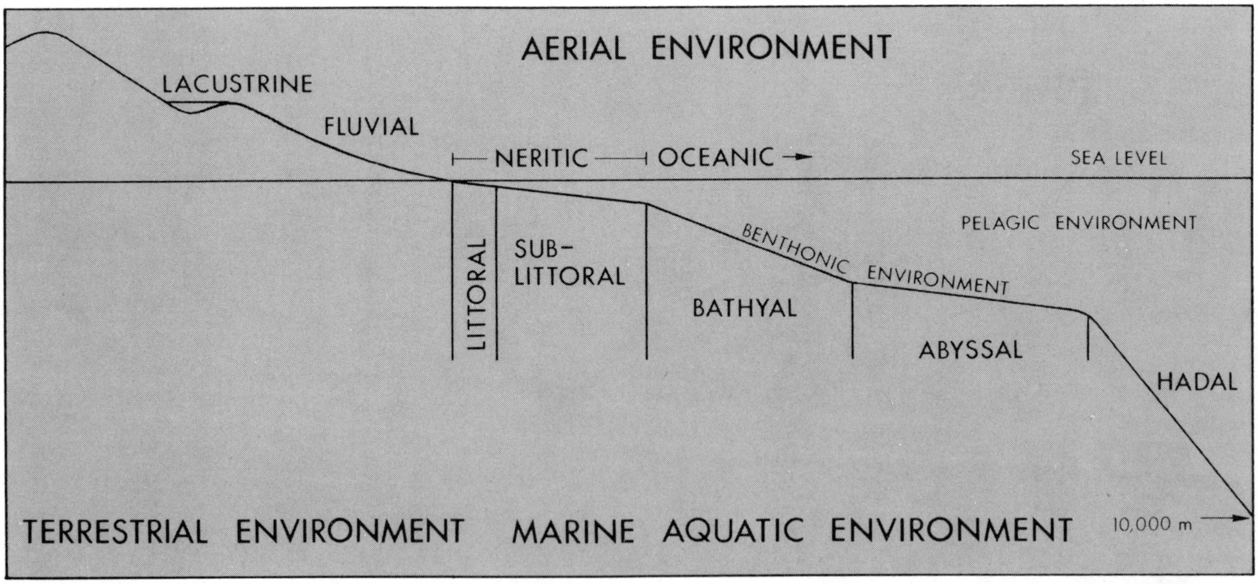

Figure 3.8. Major environments occupied by organisms.

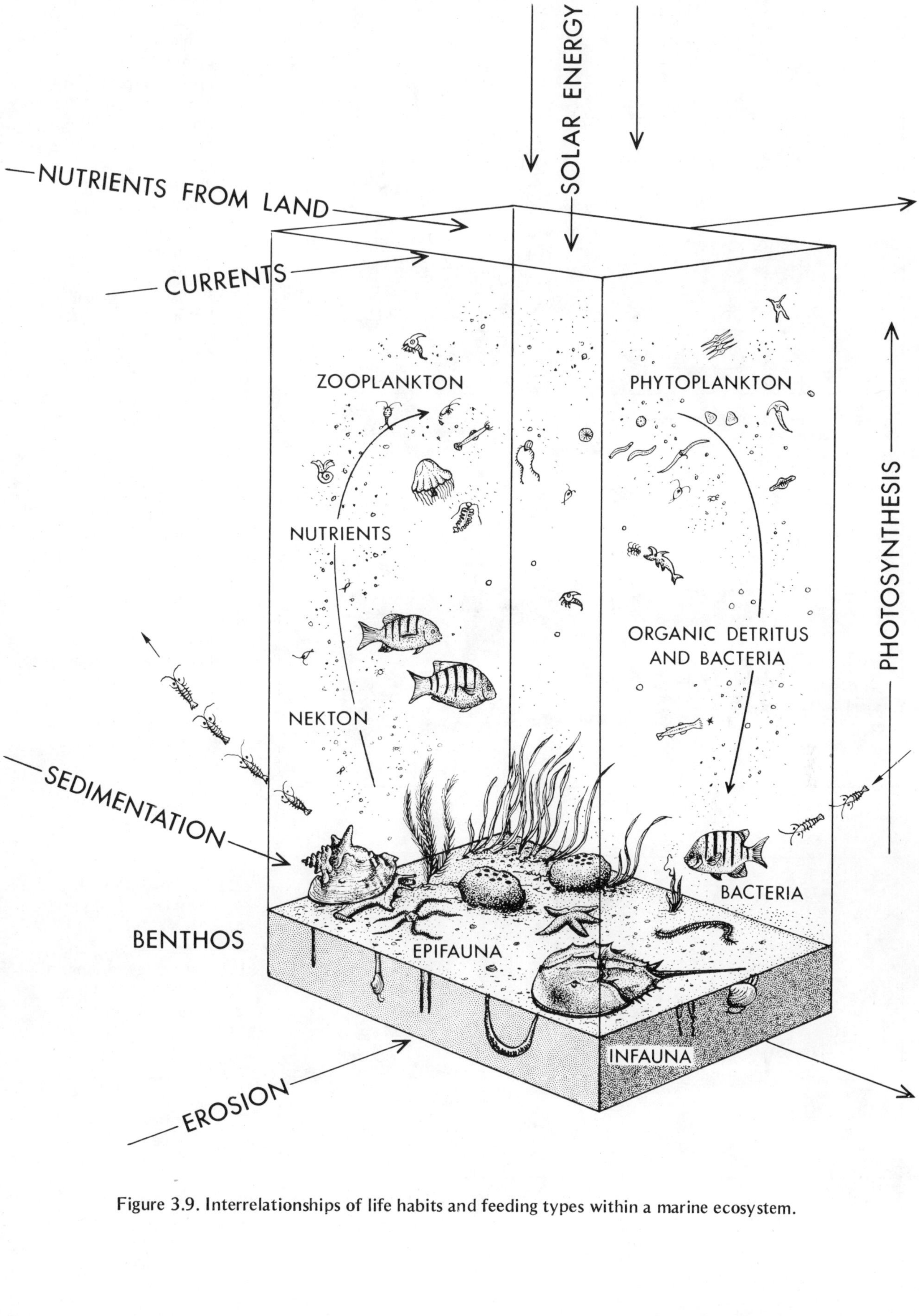

Figure 3.9. Interrelationships of life habits and feeding types within a marine ecosystem.

1. Algae and cyanophyta
2. Bacteria
3. Fungi (mold)
4. Plantae (angiosperm)
5. Protozoa (plankton)
6. Porifera (sponge)
7. Cnidaria (hydroid)
8. Mollusca (clam)
9. Mollusca (snail)
10. "Worms" (annelid)
11. "Worms" (roundworm)
12. "Worms" (flatworm)
13. Arthropoda (crayfish)
14. Arthropoda (mite)
15. Arthropoda (insect)
16. Ectoprocta (bryozoan)
17. Chordata (fish)
18. Chordata (amphibian)
19. Chordata (reptile)
20. Chordata (bird)

Figure 3.10. Freshwater organisms.

1. Plantae (moss)
2. Fungi (lichen)
3. Fungi (mushroom)
4. "Worms" (annelid)
5. Arthropoda (spider)
6. Arthropoda (insect)
7. Arthropoda (millipede)
8. Arthropoda (crab)
9. Mollusca (snail)
10. Chordata (amphibian)
11. Chordata (reptile)

Figure 3.11. Terrestrial organisms.

1. Plantae (fern)
2. Plantae (horsetail)
3. Plantae (psilophyte)
4. Plantae (lycopod)
5. Plantae (angiosperm)
6. Plantae (gymnosperm)
7. Chordata (mammal)
8. Chordata (bird)

Figure 3.12. Terrestrial organisms.

1. Chordata (reptile)
2. Chordata (bird)
3. Chordata (mammal)
4. Arthropoda (insect)

Figure 3.13. Aerial organisms.

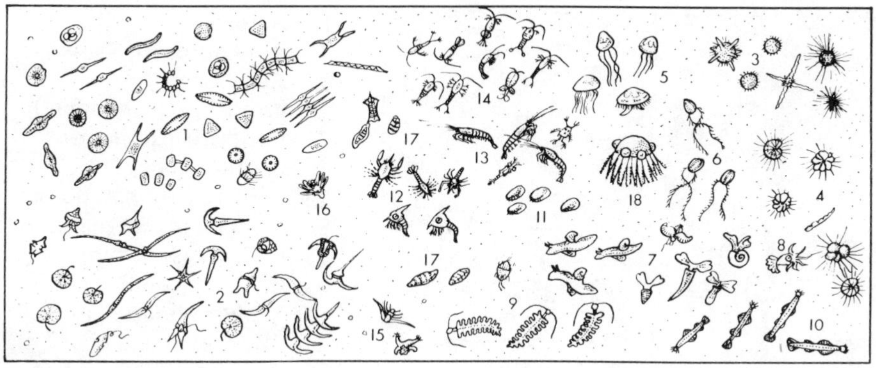

Figure 3.14. Marine plankton.

1. Algae (diatoms)
2. Algae (flagellates)
3. Protozoa (radiolarians)
4. Protozoa (foraminifera)
5. Cnidaria (medusae)
6. Ctenophora (comb-jellyfish)
7. Mollusca (snails)
8. Mollusca (nautiloid)
9. "Worms" (annelid)
10. "Worms" (arrowworm)
11. Arthropoda (ostracods)
12. Arthropoda (crab larvae)
13. Arthropoda (krill)
14. Arthropoda (copepods)
15. Echinodermata (starfish larvae)
16. Echinodermata (sea urchin larvae)
17. Chordata (tunicates)
18. Chordata (graptolite)

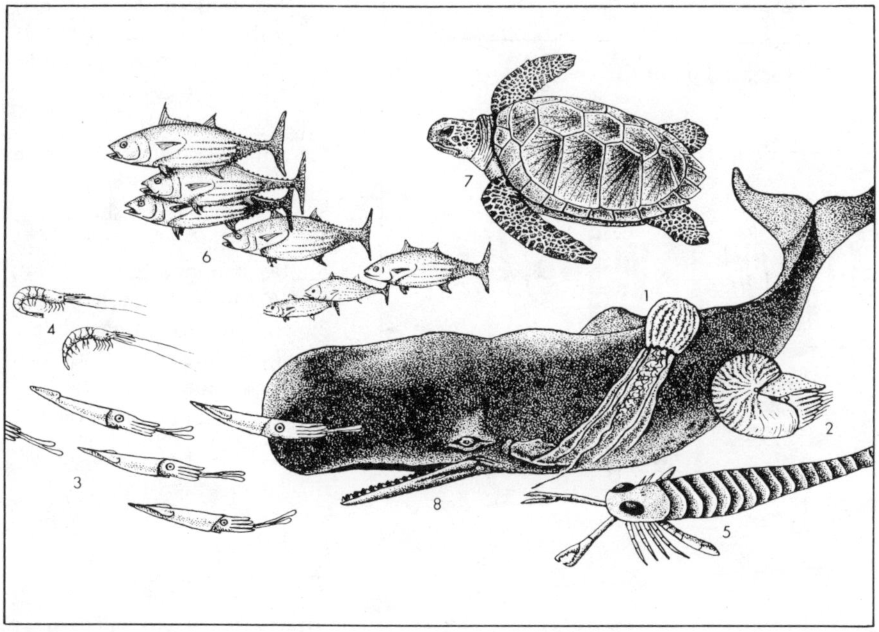

Figure 3.15. Marine nekton.

1. Cnidaria (jellyfish)
2. Mollusca (nautiloid)
3. Mollusca (squids)
4. Arthropoda (shrimp)
5. Arthropoda (eurypterid)
6. Chordata (fish)
7. Chordata (reptile)
8. Chordata (mammal)

Figure 3.16. Marine vagrant benthos.

1. Cnidaria (anemones)
2. Mollusca (snails)
3. Mollusca (chiton)
4. Mollusca (scallop)
5. Mollusca (octopus)
6. Mollusca (nautiloid)
7. Mollusca (ammonite)
8. "Worms" (ribbon worm)
9. "Worms" (annelid)
10. Arthropoda (trilobite)
11. Arthropoda (crab)
12. Arthropoda (eurypterid)
13. Echinodermata (starfish)
14. Echinodermata (sea cucumber)
15. Echinodermata (sea urchin)
16. Echinodermata (brittle star)
17. Chordata (fish)

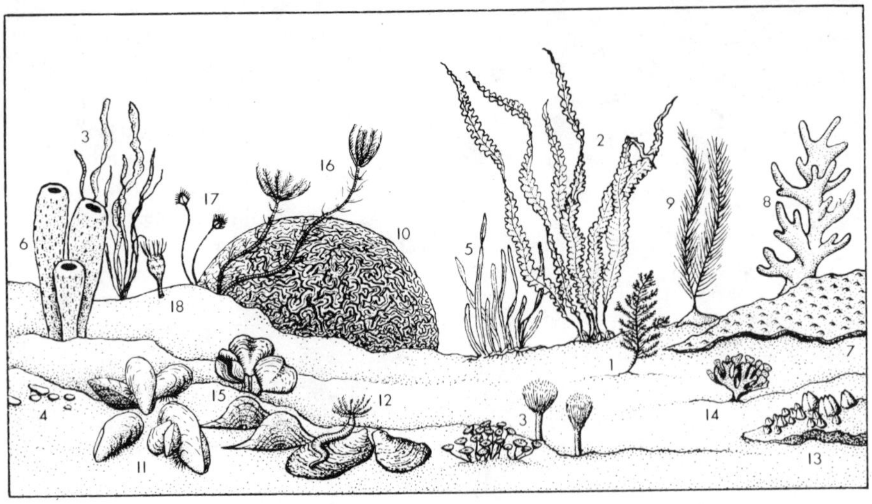

1. Algae (red)
2. Algae (brown)
3. Algae (green)
4. Protozoa (foraminifera)
5. Plantae (angiosperm)
6. Porifera (sponge)
7. Porifera (stromatoporoid)
8. Cnidaria (hydrozoan coral)
9. Cnidaria (sea whip)
10. Cnidaria (coral)
11. Mollusca (mussels)
12. "Worms" (annelid)
13. Arthropoda (barnacles)
14. Ectoprocta (bryozoan)
15. Brachiopoda (lampshells)
16. Echinodermata (crinoids)
17. Echinodermata (blastoids)
18. Echinodermata (cystoid)

Figure 3.17. Marine sessile epifauna and associated plant life.

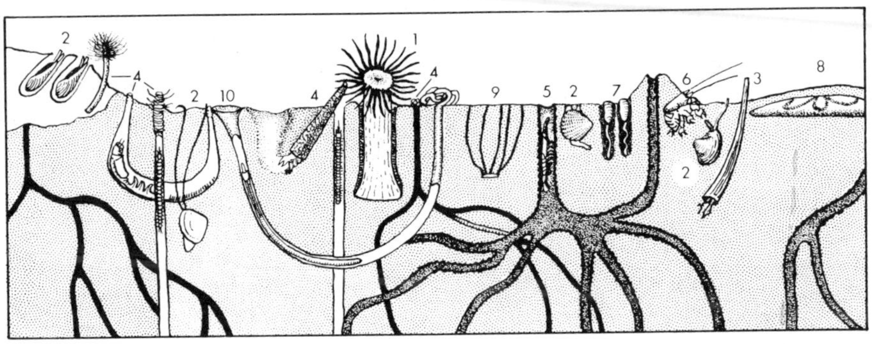

1. Cnidaria (burrowing anemone)
2. Mollusca (clams)
3. Mollusca (tusk shell)
4. "Worms" (annelids)
5. Arthropoda (ghost shrimp)
6. Arthropoda (mole crab)
7. Brachiopoda (inarticulates)
8. Echinodermata (sand dollar)
9. Echinodermata (brittle star)
10. Chordata (acorn worm)

Figure 3.18. Marine infauna.

Exercise

Several large slabs bearing natural assemblages of marine fossils have been set out in the laboratory.

1. Using the identification key, identify all of the fossil types present in each slab under a low-power microscope to find small forms.
2. Determine the feeding types and life habits of each fossil identified, using all of the information supplied in this chapter. Determine the relative proportions of each feeding type and life habit on each slab. Classify filterers into categories based on height above the substratum, if there is any indication that such diversity was present.
3. Rank all of the assemblages according to the inferred stability of the suspended food supply. If the assemblages are in detrital rocks, use the information in Chapters 1 and 2 to specify the depositional environment as closely as possible. Specify for each slab the possible causes of the stability or instability of its former ecosystem.

Questions for Discussion

1. Which fossil groups have only one feeding type and habitat? Which have more than one? What difference in shell form did you observe among fossils of the same organic group that had different life habits?

2. What are the life habits of the lower invertebrates? Is there any possible connection between the development of the coelom and the exploitation of different food supplies?
3. What features of a rock slab other than those discussed in this chapter might indicate resource stability? Could the existence of any organisms be inferred other than those that left preserved shells or skeletons?

Background Reading

Cowen, R. 1976. History of Life, pp. 23-32. New York: McGraw-Hill.
Laporte, L. F. 1968. Ancient Environments, pp. 30-60. Englewood Cliffs, N.J.: Prentice-Hall.
Levin, H. L. 1975. Life through Time, pp. 53-75. Dubuque, Ia.: William C. Brown.
McAlester, A. L. 1977. The History of Life, pp. 45-68. Englewood Cliffs, N.J.: Prentice-Hall.

Chapter Four

Geochronology
Geologic Age Dating

Objectives

1. To learn how the relative age of a rock may be known by using assemblages of fossils.
2. To learn how the absolute age of a rock may be known by using the radioactive decay of unstable elements.

Important Terms

Correlation—The determination of time relationships within an undated body of rock by relating it to a dated sequence, using either radiometric dating or fossil types.

Eon—The largest division of geologic time. The rocks deposited during an eon are termed an *eonothem.* The *Phanerozoic* eon comprises the last 600 million years of earth history and represents the time in which multicellular plants and animals are common as fossils.

Epoch—The largest subdivision of a period. The rocks deposited during an epoch are termed a *series.*

Era—The largest subdivision of an eon. The rocks deposited during an era are termed an *erathem.* The Phanerozoic eon consists of the *Paleozoic, Mesozoic,* and *Cenozoic* eras (Figure 3.1).

Nuclide—A type of atom characterized by the makeup of its nucleus. Unstable nuclides are involved in radioactive dating.

Period—The largest subdivision of an era. The rocks deposited during a period are termed a *system.* The Phanerozoic periods are listed in Figure 4.1.

Introduction

The determination of geologic time presents one of the most difficult problems in geology. In some areas excellent time markers, such as beds of volcanic ash, are locally extensive and allow the geologist to correlate the rocks through which the time marker passes. Such features are relatively rare, and they usually do not aid in correlation across an entire continent or from one continent to another. When igneous rocks cut across sedimentary rocks or are buried by them (for example, dikes, sills, or lava flows), the radiometric age of the igneous rock gives us either an upper or a lower

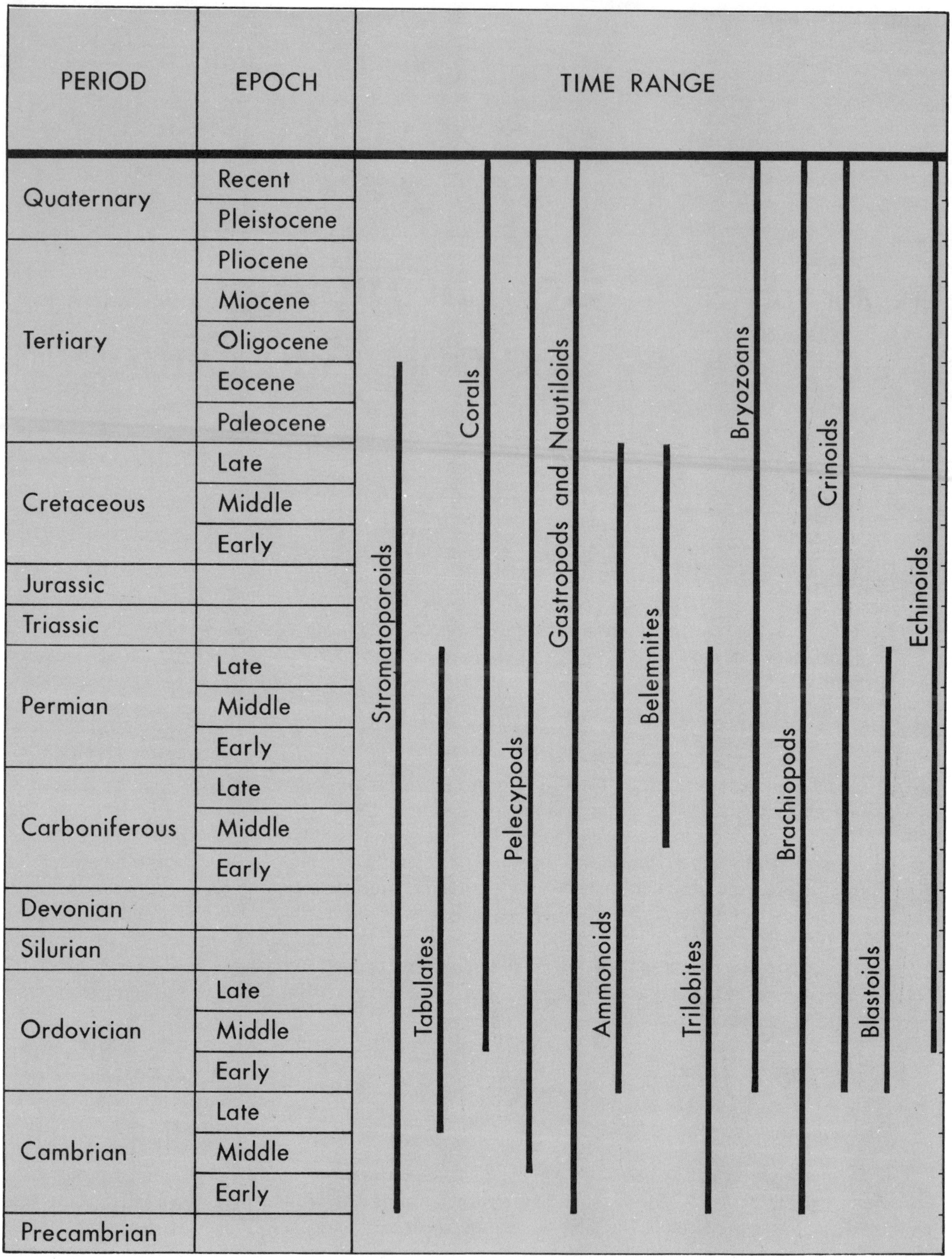

Figure 4.1. Time ranges of fossil groups identified in Chapter 3.

age limit for the sedimentary rocks, but not a precise age. Geologically young sediments can be dated directly using ^{14}C (back to 40,000 years) and ^{230}Th (back to 300,000 years). Glauconite, a mineral silicate of potassium, aluminum, and iron, often crystallizes within marine sediments at their time of deposition; a direct radiometric age date for the sediment can be found by measuring the decay of unstable potassium to argon. Unfortunately, many sediments do not contain glauconite, and in those that do, the leakage of argon gas with time causes the age dates to be too recent. Most often, however, these direct methods of age determination (explained in detail in the second part of this chapter) cannot be used, and the geologist must use indirect methods, primarily using fossils.

Physical stratigraphy, the tracing out of beds of uniform rock type across a large area, is of very little value in determining time relationships, because a homogeneous rock unit usually represents a single depositional environment that has migrated spatially, so that the same rock unit represents deposition at different times in different geographic areas. In fact, a greater duration of time may be present in the lateral development of a homogeneous rock unit across a large area than in the vertical sequence of superposed rock units at a single locality. For a simple example, a very large body of sandstone in the western United States is of late Precambrian age in Nevada and becomes progressively younger as it is traced eastward until, in the Dakotas, it is Early Ordovician, thus spanning a time interval of more than 100 million years. The vertical thickness of the sandstone in any one place probably represents less than 1 percent of the time represented laterally. The types of stratigraphic simulation models used in Chapters 13 through 16 may be very useful in determining the time relationships within large bodies of sedimentary rock.

Age Determination from Fossils

Because of the general absence of any fossils, especially of multicellular organisms, time correlation is extremely difficult in the eonothems deposited before the Phanerozoic. Only the Phanerozoic eonothem contains abundant fossils of animal life, and their time relationships are used for dating unknown sequences. Figure 4.1 shows the known time ranges of the fossil types identified in Chapter 3. The presence of a single type of fossil, such as gastropods, does not clearly specify any time period; however, the presence of several fossil types together in the same rock unit brackets the time period more closely. For example, a rock containing ammonoids, belemnites, and trilobites could only have been deposited during the interval from the Middle Carboniferous to the Late Permian. This method of time determination is known as the *overlapping range method.* The precision of this method depends upon the accuracy of the fossil identifications as well as accurate knowledge of their time ranges. The animals listed in Figure 4.1 are identified only to the phylum or class level, and thus their time ranges are very long. The more precisely a fossil is identified, from order and family down to genus and species, the narrower are the time limits that it specifies. Most species have a range less than one epoch; genera, less than a period; families, less than an era. Orders, classes, and phyla have much longer durations in geologic time.

Exercise I

1. Using the overlapping range method and the time ranges given in Figure 4.1, identify as closely as possible the age of each of the unknown slabs set out in the laboratory. Each slab will contain several different fossil groups, and its age must be within the time interval in which the ranges of these groups overlap.
2. As an optional exercise, identify all of the fossils in each slab down to the level of *family.* To do this, you will need special help from your instructor or specially prepared identification keys designed for the materials you are working with. The book *The Fossil Record* (edited by

W. B. Harland et al.) lists the time ranges for all families of fossil organisms as precisely as possible insofar as they are known. Using the time ranges given in *The Fossil Record,* determine the age of the unknown slabs as closely as possible. Compare the results with those obtained in the first part of the exercise.

Radiometric Dating

The radioactive decay of certain unstable elements (for example, ^{14}C, ^{40}K, ^{238}U) provides the most precise geologic clocks and has been the basis for the current belief in the tremendous length of geologic time. Measuring the age of rocks and minerals by radioactive decay began in 1907, when it was found that the ratio of daughter lead to parent uranium was constant in rocks of the same age. Later developments in geochronology showed that meteorites that fall to earth were all formed some 4.4 to 4.7 billion years ago, during the early history of the solar system; that rocks of the moon's crust are 3.5 to 4.2 billion years old; and that the oldest remaining portions of the earth's crust are 3.5 to 3.8 billion years old.

Radioactive decay is the property some nuclides have of transforming spontaneously into other nuclides by emitting charged particles from the nucleus. Because this process is statistically random, the *probability* that a nucleus will decay in any given time interval is the same for all nuclei of a particular radioactive nuclide. The *rate* at which a radioactive nuclide decays does not depend on the physical or chemical state of the sample, only on the *kind* and *number* of nuclei present.

The probability of decay of a radioactive nuclide can be expressed by a decay constant (λ), which expresses the fraction of nuclei that decay per unit of time. If there are N nuclei present, then the total number of decays per unit of time is λN. This rate is given by

$$\frac{dN}{dt} = -\lambda N \qquad (4.1)$$

The negative sign indicates a decrease in the number of nuclei with time. For an observation over some time interval $t_0 \rightarrow t$, during which the number of nuclei remaining changes from $N_0 \rightarrow N_t$, we can derive an expression for t by rearranging equation 4.1,

$$\frac{dN}{N} = -\lambda dt$$

and integrating it.

$$\int_{N_0}^{N_t} \frac{dN}{N} = \int_{t_0}^{t} -\lambda dt$$

If $t_0 = 0$, then integration gives

$$\ln N_t - \ln N_0 = -\lambda t$$

$$\ln \left(\frac{N_t}{N_0}\right) = -\lambda t$$

$$\frac{N_t}{N_0} = e^{-\lambda t}$$

This can be rearranged to

$$N_t = N_0 e^{-\lambda t} \qquad (4.2)$$

Equation 4.2 is the *general decay law* and indicates that if we can measure N_0 and N_t for some geologic sample, then we can calculate the sample's age.

It is not usually possible to measure N_0, but it is quite easy to determine the amount of the decay product or daughter (D) at time t, which will permit determination of t. This is so because the number of parent nuclei originally present must equal the number of parent nuclei at some later time t plus the number of daughter nuclei that have been produced during that time. That is,

$$N_0 = N_t + D_t$$

If we substitute for N_0 from the general decay law, then

$$N_t + D_t = \frac{N_t}{e^{-\lambda t}}$$

$$N_t + D_t = N_t e^{\lambda t}$$

and rearranging gives

$$D_t = N_t e^{\lambda t} - N_t$$

$$D_t = N_t(e^{\lambda t} - 1) \qquad (4.3)$$

Therefore, if we measure the amount of parent and daughter nuclei at time t we can determine the age of a rock or mineral.

A particularly useful concept in dealing with radioactive decay is the *half-life* ($t_{1/2}$), the time required for one-half the original number of nuclei to decay (see Figure 4.2). The half-life is related to the decay constant as follows.

$$\ln\left(\frac{N_t}{N_0}\right) = -\lambda t$$

At $t_{1/2}$,

$$N_t = \tfrac{1}{2}N_0$$

$$\frac{N_t}{N_0} = \tfrac{1}{2}$$

Thus,

$$\ln \tfrac{1}{2} = -\lambda t_{1/2}$$

and

$$t_{1/2} = \frac{-\ln(\tfrac{1}{2})}{\lambda}$$

Now

$$-\ln(\tfrac{1}{2}) = \ln\left(\frac{1}{\tfrac{1}{2}}\right) = \ln 2$$

so that

$$t_{1/2} = \frac{0.693}{\lambda}$$

In geochronology the limit of any dating method is usually about seven half-lives. Thus for ^{14}C, with a half-life of 5700 years, the oldest material that can be dated is about 40,000 years old. Common radioactive decay schemes in geology are shown in Table 4.1.

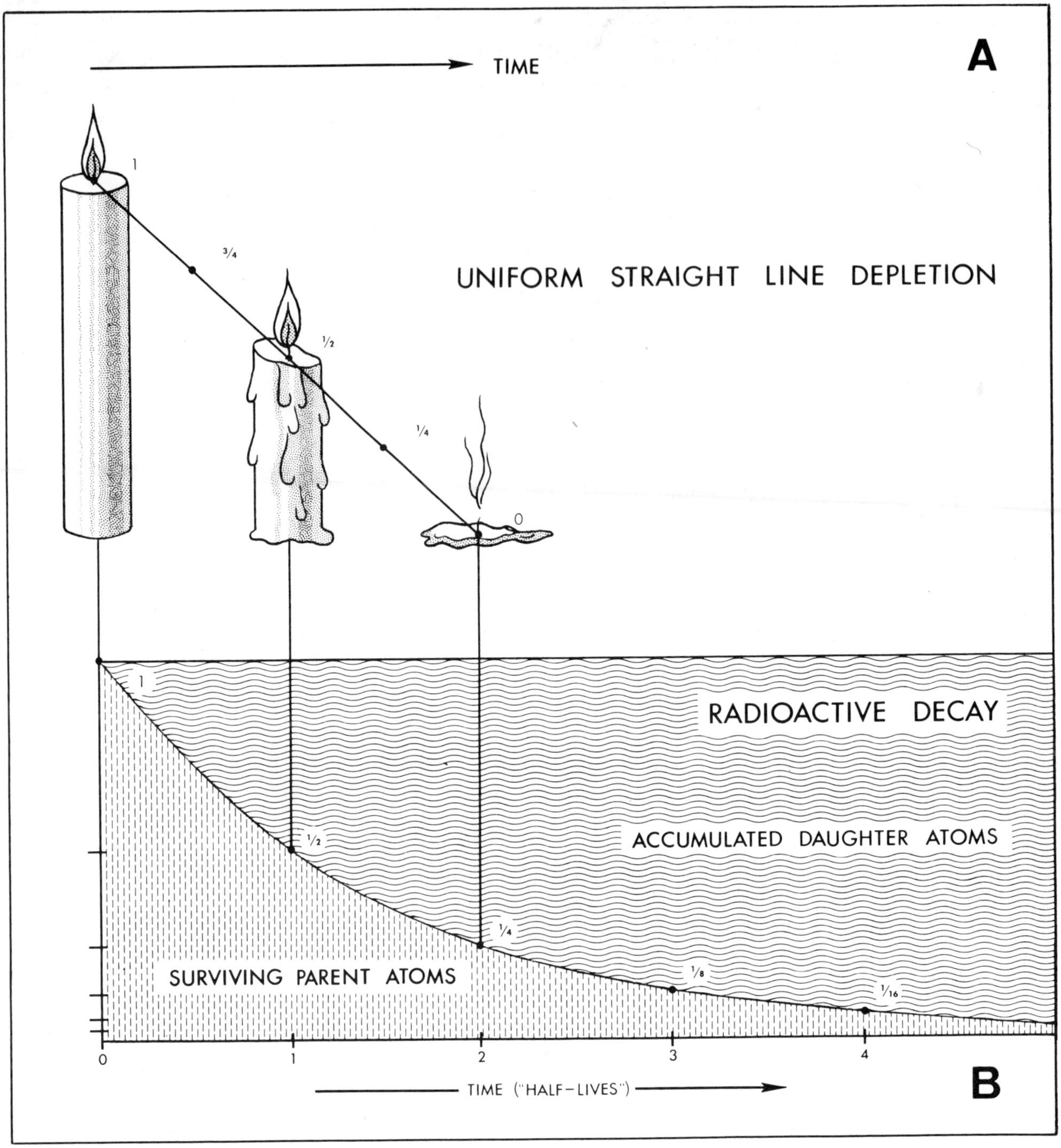

Figure 4.2. Uniform depletion process (A), contrasted with radioactive decay (B). Note that the radioactive decay curve approaches the zero line asymptotically and that the end of one half-life interval is the beginning of a new one.

TABLE 4.1. Common Radioactive Decay Schemes

System		λ *(Per Year)*	$t_{1/2}$ *(Years)*
$^{14}C \rightarrow {}^{15}N$		1.209×10^{-4}	5730
$^{40}K \rightarrow {}^{40}Ar$		0.585×10^{-11}	
$\rightarrow {}^{40}Ca$		4.72×10^{-10}	
	λ total =	5.305×10^{-10}	1.31×10^{9}
$^{87}Rb \rightarrow {}^{87}Sr$		1.39×10^{-11}	5.0×10^{10}
$^{238}U \rightarrow {}^{206}Pb$		1.54×10^{-10}	4.5×10^{9}
$^{235}U \rightarrow {}^{207}Pb$		9.7×10^{-10}	7.13×10^{8}
$^{232}U \rightarrow {}^{208}Pb$		4.99×10^{-11}	1.41×10^{10}

Sample Problems

1. What is the decay rate (number of charged particles emitted per unit time) of 1 gram of ^{238}U?
   a. The gram molecular weight of ^{238}U is 238 grams/mole.
   b. $\lambda^{238}U = 1.54 \times 10^{-10}$ year.
   c. 1 second = 3.17×10^{-8} year.
   d. Avogadro's number = 6.023×10^{23} atoms/mole.

Solution:

$$\frac{-dN}{dt} = -\lambda N$$

$$\frac{dN}{dt} = \left(\frac{1.54 \times 10^{-10}}{\text{year}}\right)\left(\frac{3.17 \times 10^{-8}\ \text{year}}{\text{second}}\right)\left(\frac{1\ \text{gram}}{238\ \text{grams/mole}}\right)\left(6.023 \times 10^{23}\ \text{atoms/mole}\right)$$

$$\frac{dN}{dt} = 12{,}300\ \text{atoms/second}$$

2. A nuclide, x, is radioactive and decays to a stable (nonradioactive) daughter nuclide, y, with a half life of 1.5×10^{8} years. You have collected a granite sample which contained no y at the time it crystallized. Your analysis shows that the whole rock contains 1×10^{12} atoms of x and 8×10^{9} atoms of y. What is the age of the granite? As a first approximation assume that $(e^{\lambda t} - 1) = \lambda t$.

   a. $t_{1/2} = \dfrac{.693}{\lambda}$

   Thus, $\lambda = \dfrac{.693}{t_{1/2}} = \dfrac{.693}{1.5 \times 10^{8}\ \text{years}} = 4.62 \times 10^{-9}/\text{year}$

   b. $D_t = N_t\,(e^{\lambda t} - 1)$ $\quad D_t = 8 \times 10^{9}$; $N_t = 1 \times 10^{12}$

   $8 \times 10^{9} = (1 \times 10^{12})(e^{\lambda t} - 1)$

   $8 \times 10^{9} = (1 \times 10^{12})(4.62 \times 10^{-9}/\text{year})(t)$

   $t = 8 \times 10^{9} / 4.6 \times 10^{3}/\text{year}$

   $t = 1.73 \times 10^{6}$ years, or 1,730,000 years old

Exercise II

You will need a calculator or slide rule and a scale. Linear and semilogarithmic graph paper will be found near the back of the book.

1. A radioactive isotope has a half-life of 10 days.
   a. What fraction (N_t/N_0) of it remains after 1 day? After 5 days? 10 days? 20 days? 50 days?
   b. Plot N_t/N_0 versus time on linear graph paper and on semilogarithmic graph paper.
2. A 1-gram sample of the mica muscovite has 2000 parts per million of rubidium. The radioactive isotope constitutes 28% of total elemental rubidium, and ^{87}Rb has a decay constant $\lambda^{87}Rb = 1.39 \times 10^{-11}$/yr. Calculate how much ^{87}Rb would be produced if the muscovite sample was 1.5×10^9 years old (Precambrian) and if it was 1×10^6 years old (Quaternary). Assume that there is no strontium in the sample and that $(e^{\lambda t} - 1) = \lambda t$.

Questions for Discussion

1. Which method provides more precision in age dating, using fossils or radioactive decay? Can the two methods be legitimately compared? List the possible sources of error in each method. Why are both methods needed in geology?
2. Recently, several organisms that were presumed to have become extinct hundreds of millions of years ago have been found living in very deep water or otherwise inaccessible environments. Among these are the coelacanth fish and the monoplacophoran molluscs. The living sclerosponges closely resemble the supposedly extinct stromatoporoids; they now live in crevices and cavities in reefs. What are the broad implications of these discoveries for geologic time correlation using fossils?
3. The genus *Hyolithellus*, a very common fossil in Early Cambrian strata, is now considered to be a fossil representative of the phylum Pogonophora, the bearded worms, close relatives of the vertebrates and a group once thought not to have a fossil record. Also, many fossils of animals without hard parts have now been found, in which the soft tissues are preserved as carbonized films or impressions in the rock. These discoveries have provided a fossil record for many groups that otherwise would have none. In this light, are the observed first appearances of fossil groups any more reliable than their last appearances (observed times of extinction)? To what extent does the *observed* time range of a fossil organism depend upon its preservability and abundance in certain rock-forming environments?

Background Reading

Berry, W. B. N. 1968. Growth of a Prehistoric Time Scale Based on Organic Evolution. San Francisco: W. H. Freeman.

Eicher, D. L. 1976. Geologic Time, 2nd ed. Englewood Cliffs, N.J.: Prentice-Hall.

Harland, W. B. et al., eds. 1967. The Fossil Record. London: Geological Society.

Raup, D. M., and Stanley, S. M. 1971. Principles of Paleontology, pp. 317-52. San Francisco: W. H. Freeman.

Chapter Five

Paleoecology Fossils as Environmental Guides

Objectives

1. To evaluate the usefulness of different kinds of fossils in interpreting the physical conditions in ancient environments.
2. To learn the ecologic tolerances and preferences of living organisms commonly preserved as fossils.
3. To consider how environmental conditions can affect an animal's growth and skeletal form and how such effects are useful in paleoecology.

Important Terms

Norm of reaction—The total variety of forms that a single type of organism can have, given different external conditions during growth.

Paleoecology—The study of factors controlling the distribution and abundance of fossil organisms in bodies of sedimentary rock.

Preference—A narrow range of conditions in which an organism functions best, with respect to physical conditions and competition with other organisms.

Salinity—The saltiness of the water, caused by dissolved mineral salts.

Tolerance—The range of conditions which an organism can withstand and survive.

Turbidity—The muddiness of the water, caused by suspended sediment.

Introduction

Discovering the exact physical conditions an ancient animal or plant lived in is true detective work, and all available clues must be considered together in solving the mystery. You will learn to use a variety of strategies.

1. Comparison of the fossil with its nearest living relatives. The chief pitfall here is that unless the fossil is that of a living *species,* which is uncommon, its ecologic functions may not be identical to its living relatives.

2. Comparison of the fossil with its fossil relatives whose paleoecology is known to some extent. The pitfall here is the same, compounded by the piling of inference on inference.
3. Comparison of the fossil with the sediments and sedimentary structures in which it is found. The chief danger in this approach is that the fossil may have been buried in an environment other than the one it lived in. A second problem is that the conditions can only be specified in a very general way.
4. Comparison of the fossil with other fossils found with it whose paleoecology is known to some extent. This approach is subject to all of the pitfalls described above.
5. Comparison of the form (*morphology)* of the fossil with similar forms in other organisms whose function or ecology is known. The first pitfall is that the form of any organism may represent a compromise between two or more functions that it carries out. The second is that there may not be a cause and effect relationship between the form of a living organism and its environment.
6. Comparison of the chemistry or mineralogy of the shell or skeleton with that of living organisms in which these features are affected by the environment. The danger here is that the chemistry of fossil shells has usually been altered by their burial and lithification in sedimentary rocks.

Analogies with Living Counterparts

Some of the best information concerning the tolerances and preferences of ancient organisms comes from living counterparts that have similar ecologic requirements. Such analogies are particularly successful in determining ancient turbidity and salinity.

The amount of sediment suspended in the water restricts the distribution of many living organisms (Figure 5.1). Notice that suspension feeders, particularly sponges, corals, bryozoans, and crinoids, are especially restricted by turbidity. Many aquatic animals filter water through their organs to obtain food and dissolved oxygen and to dissipate waste materials. Many species have special adaptations such as pore systems, sieve plates, meshes, and spines to keep out particles that are too large. Smaller particles, especially silt and clay, may interfere with feeding or respiration. Notice also that certain types of algae are very sensitive to turbid water. Is it for the same reason?

Turbidity, as you learned in Chapter 1, is related to the rate of deposition of suspended sediment. Only infaunal organisms can stand very rapid deposition, because they normally live in sediment and are not particularly bothered by a little more. In epifaunal animals, the shell form may tell you something about rates of deposition. If deposition is too rapid, certain organisms, especially those attached to the bottom, may not be able to grow or burrow up through the covering sediments quickly enough to avoid burial. Storms can easily kill off the bottom fauna in normally clear water areas by burying them with sediment that settles out of the water when the storm subsides. In general, low, flat, thin shells or colonies are much more susceptible to burial than tall erect or branching forms. Response to turbidity and sedimentation can be inferred, then, from (1) living counterparts and (2) the morphology of the shells or skeletons.

The distribution of many organisms is greatly influenced by the amount of dissolved salts in the water (Figure 5.2), a factor which distinguishes several major aquatic environments. *Fresh* water is found only on land masses. *Brackish* water is found in bays, lagoons, estuaries, deltas, and river mouths, all of which are marine shoreline environments. *Marine* salinities (30.0 to 40.0 parts per thousand of dissolved salt) are characteristic of the open oceans and the continental shelf waters up to the shoreline. *Hypersaline* conditions characterize some coastal lagoons and lakes in arid regions that have little circulation or no external drainage.

Calcareous-shelled brachiopods, stony bryozoans, corals, echinoderms, and cephalopods are good indicators of marine salinities. Echinoderms cannot tolerate even slight changes in salinity, because seawater fills their water vascular system. Abundant oysters or phosphatic brachiopods are

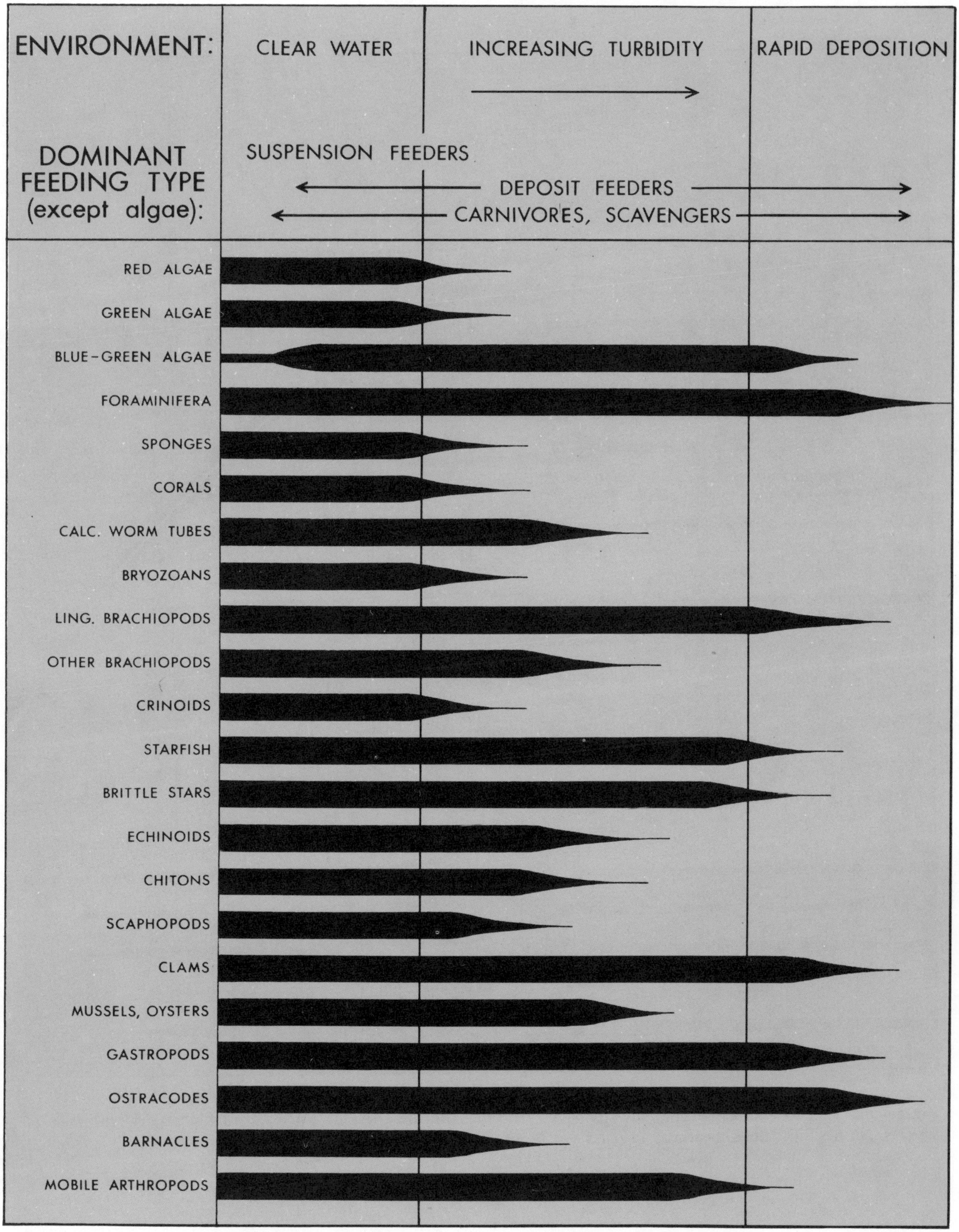

Figure 5.1. Response of organisms now living to turbidity and sedimentation.

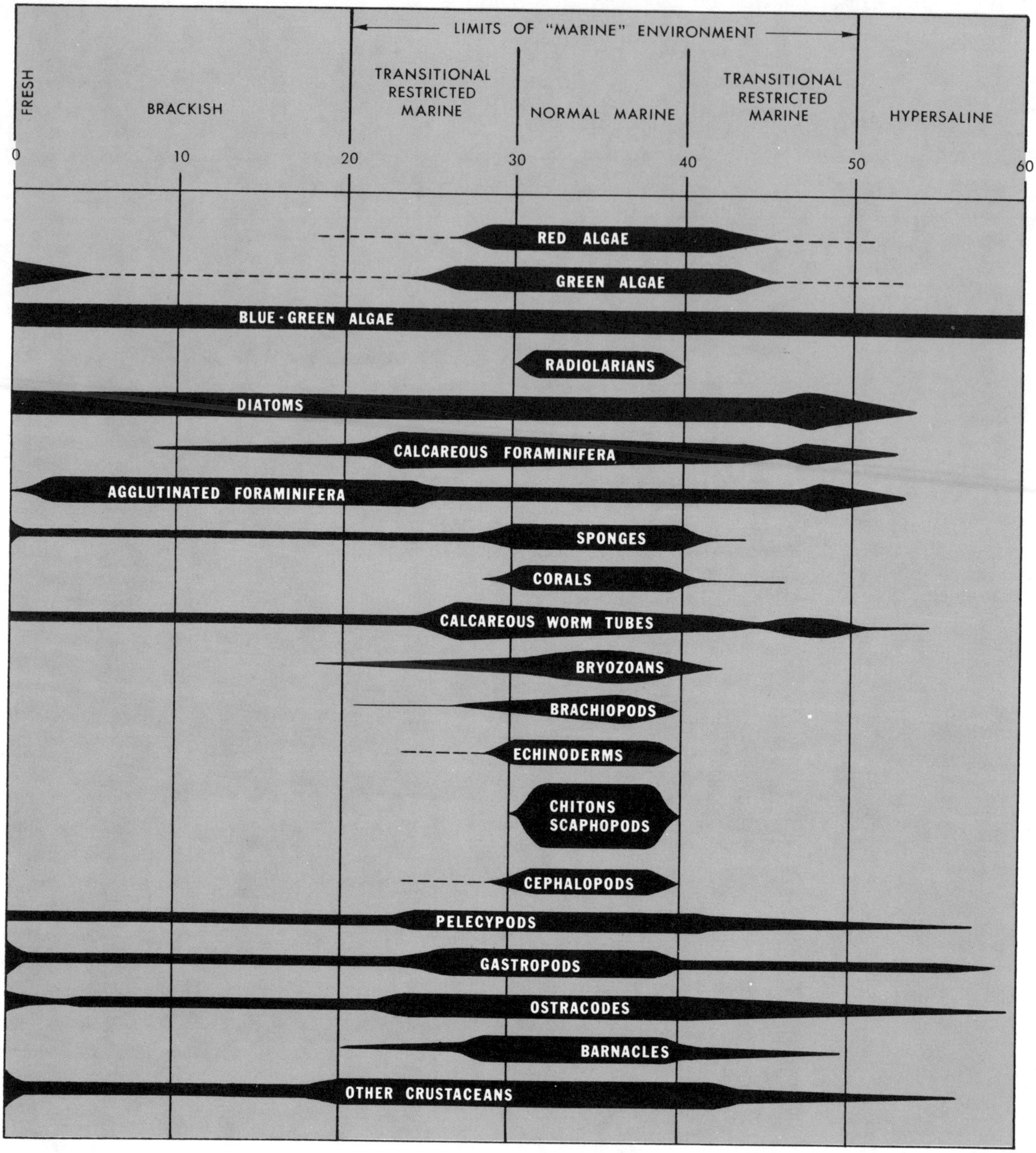

Figure 5.2. Response of organisms now living to salinity. The thickness of the bar roughly reflects numbers of species. Numerical scale is measured in parts per thousand of dissolved salt.

good indicators of brackish water (0.5 to 30.0 parts per thousand), particularly if normal marine animals are absent. Negative evidence is just as important in determining ancient environmental conditions as positive evidence. Abnormal shell growth may occur in some organisms if the salinity is unusually high or low.

Norms of Reaction

Organisms can develop in slightly different ways when growing under different external conditions. Food supply, light, oxygen, and water movement can all affect the growth of marine organisms. Geneticists refer to such changes as the norm of reaction, or the amount of morphologic leeway that the genetic makeup of the organism will permit. Such variation might cause problems for a taxonomist, but it may be very helpful for understanding an ancient environment. Just as sediment grains respond to different levels of water movement, so do the forms of many of the organisms living in the water. Water movement affects growth in many ways—by stresses on the growing animal, by irritation of the shell-forming tissues, by turbidity and sedimentation around the animal, and by causing differences in the supply of food and oxygen and in the removal of waste materials.

Less is known of the norms of reaction of solitary organisms than of colonies. Living brachiopods become wider and more ornamented in deep, quiet water, but fatter and smoother shelled in shallow, agitated water. Living mussels (clams) have shorter and rounder shells in rough water, and longer, more angular shells in quiet water. Both groups tend toward more spherical, deep-bodied forms with thicker shells in more agitated conditions, and thinner, shallow-bodied forms with thinner shells in quieter water.

More is known about norms of reaction in colonial animals, such as corals, bryozoans, and sponges. Environmental effects on colony form as well as on individuals within the colony are shown for two modern corals (Figures 5.3 and 5.4). In general, shorter, stockier forms inhabit more agitated water, and taller, branching forms live in quieter water. These effects result from the need to escape from encroaching sediment in quiet water and to offer less resistance to water flow in moving water. Living corals often grow into very wide and flat colonies in deep water because they have symbiotic algae that need light for photosynthesis; the colony grows into a flat disk to get the maximum exposure to light at depth. Paleozoic corals, however, did not have symbiotic algae and consequently did not respond in this way to water depth.

Bryozoan colonies also tend to be flat or encrusting in agitated water, and to grow into erect and branching colonies in quiet water. The Ordovician bryozoan *Dekayia* (Figure 5.5) shows colonial forms similar to those of some corals in beds of different lithology. In living species of bryozoans on the continental shelf, the percentage of erect colonies increases as the water becomes deeper (Figure 5.6). This relationship is generally true of most colonial organisms, because deeper water is usually less agitated than shallow water; the major exception is found in colonies depending on photosynthesis. The encrusting protozoan *Homotrema rubrum* shows a similar pattern (Figure 5.7). In rough water all forms appear because quiet water conditions exist in small crevices and underneath shells and large corals, but in completely quiet water only the branching forms are found.

Exercise

Several large slabs bearing assemblages of marine fossils will be set out in the laboratory.

1. Using analogies with living counterparts and growth forms, determine as closely as possible the turbidity and salinity of the macrohabitat that each slab represents. Be sure to use negative evidence as well. Your instructor might have to provide special help to identify some of the fossils.

1. Delicate branching form from below low tide level, on coral knolls in protected lagoons and other sheltered localities.

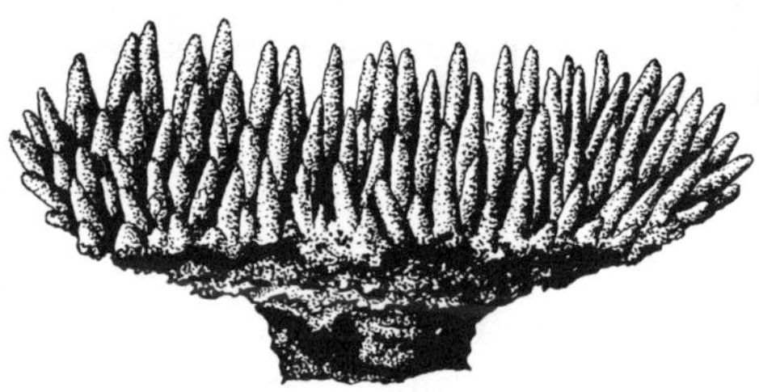

2. Shorter, stumpy, cylindrically branched form from water only a few meters deep in protected sites.

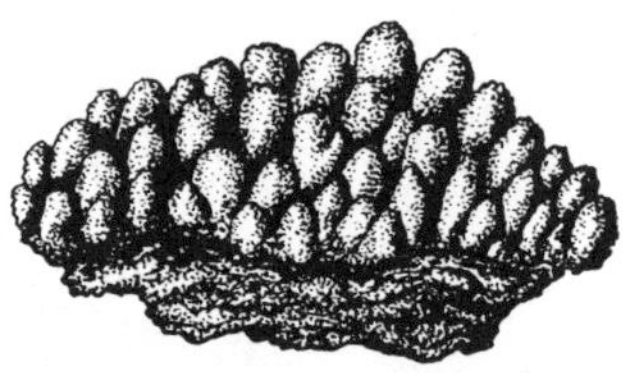

3. Broad, almost encrusting form with short, closely packed polygonal branches from areas exposed to rough water and only inches deep at low tide.

1. Tall, erect form from deep water.

2. Intermediate, stocky form with conelike accordant crests common at shallower depths.

3. Massive, very stocky form from shallows near the reef crest.

A. Three growth forms of *Acropora humillis* from the Marshall Islands modified by differences in water turbulence and depth.

B. Three growth forms of *Acropora palmata* from Andros Island in the Bahamas modified by differences in water turbulence.

Figure 5.3. Environmental influence on the coral genus *Acropora*.

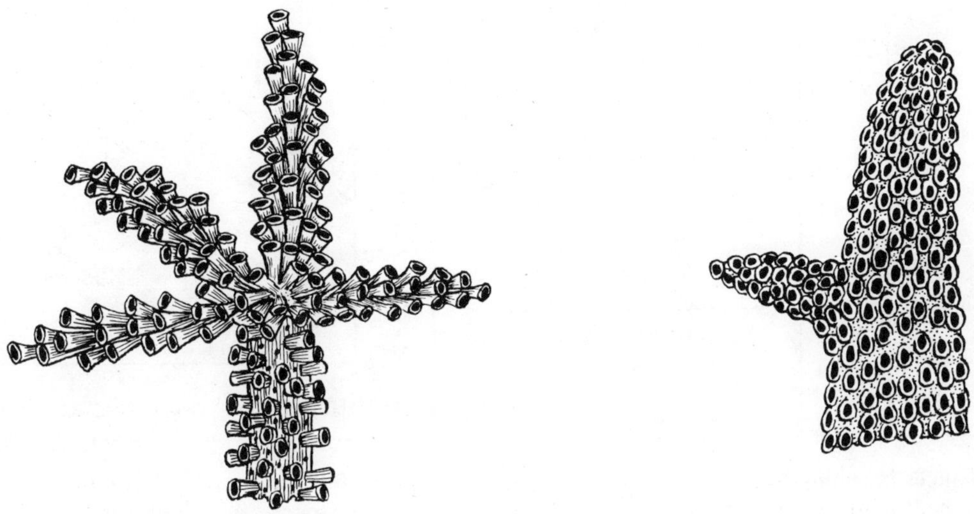

1. Corallites raised above surface of colony in turbid, sediment-laden water.

2. Corallites flush with surface of colony in clean, sediment-free water.

C. Differences in corallites of *Acropora pulchra* from the Pacific, demonstrating a response to sedimentation.

Figure 5.3 continued

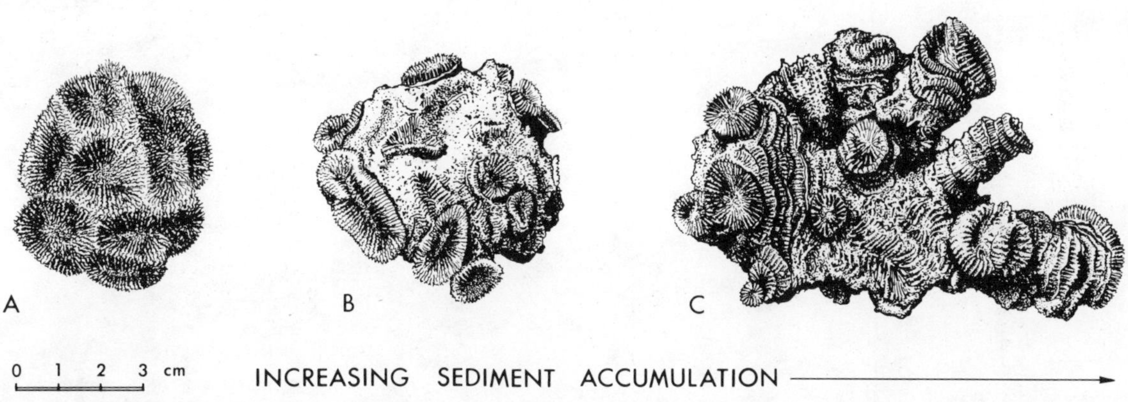

Figure 5.4. Variations in growth form of the coral *Isophyllia multiflora* from Florida Bay, Florida. (A) Specimen from a mud-free rocky platform showing "normal" development of colony form. (B and C) Specimens from a carbonate mud bank showing effects on growth form with increased sediment accumulation. Columnar projections allow the coral to elevate its living tissue above encroaching mud.

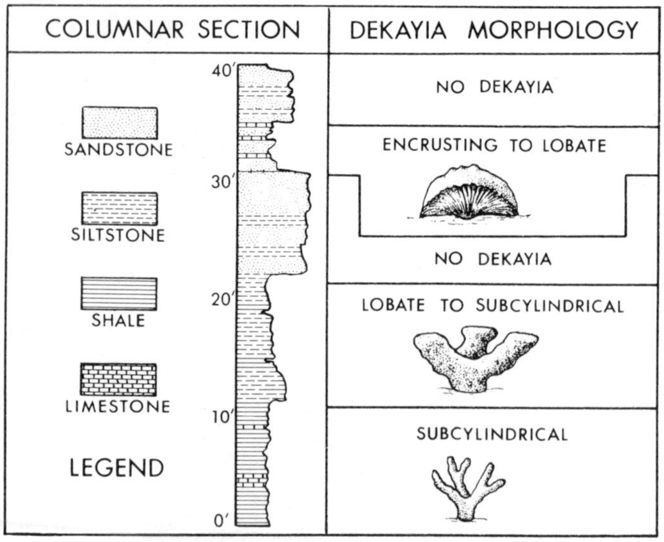

Figure 5.5. Changes in colony form in the bryozoan *Dekayia* associated with beds of differing lithology.

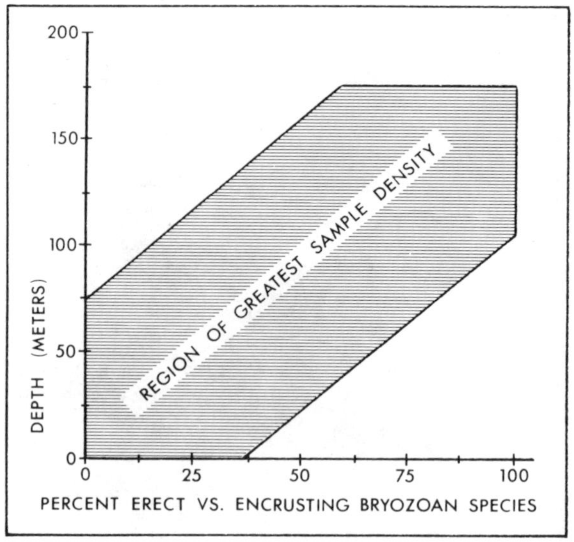

Figure 5.6. Increase of erect colony forms of bryozoan species with increasing water depth on the continental shelf off New England.

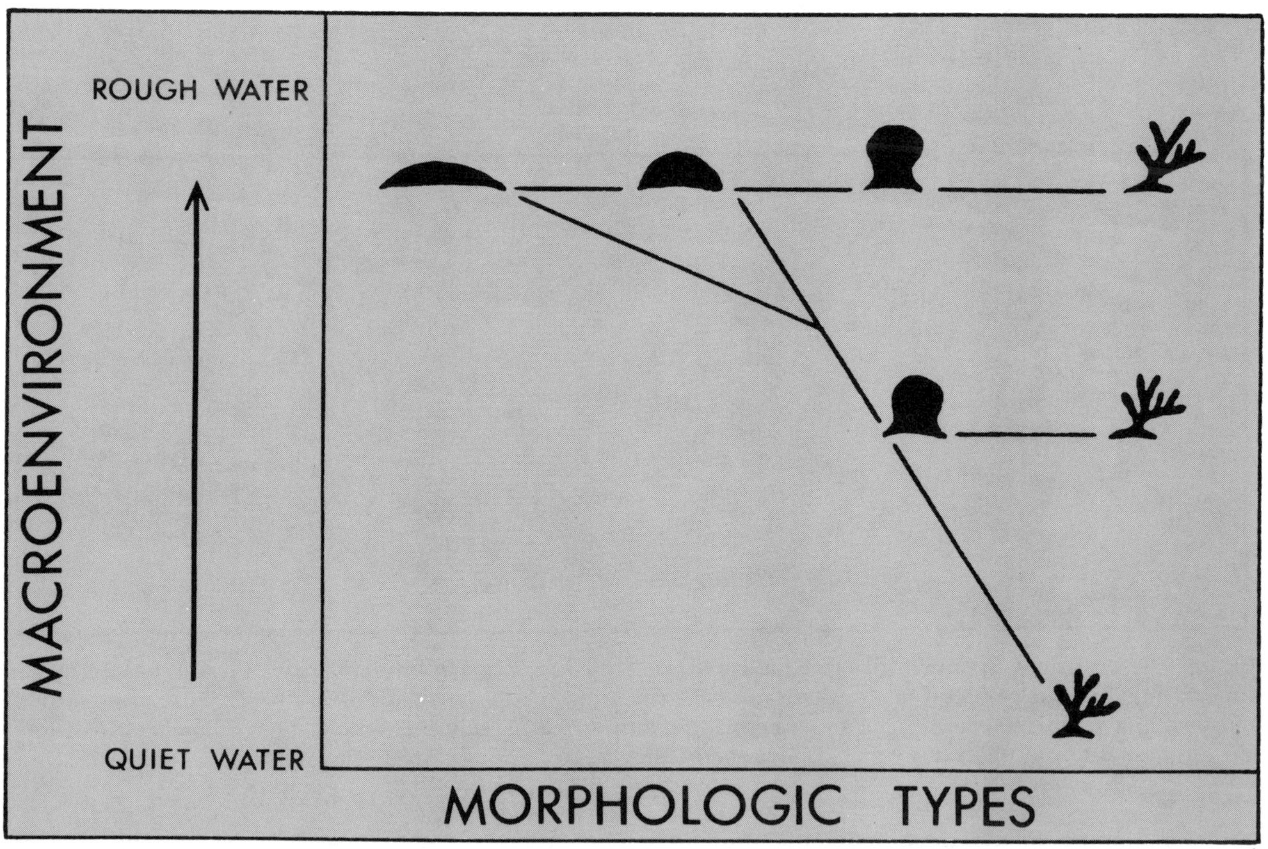

Figure 5.7. Growth forms of the encrusting protozoan *Homotrema* resulting from different degrees of water turbulence.

2. Using the norms of reaction that were described for shell and colony form, determine the water agitation and possible depth for each slab. Are the organic growth forms in each case consistent with the sediment characteristics?

Questions for Discussion

1. In which colonial forms should the individuals be more closely packed in the colony? Will the closeness of packing affect the shapes of the individuals within the colony? Could you possibly identify the overall colony form from just a small fragment?
2. There is a general relationship between degree of water agitation and depth. In what environments could this relationship be inverted? What additional evidence could verify such an inversion?

Background Reading

Heckel, P. H. 1972. "Recognition of Ancient Shallow Marine Environments." In J. K. Rigby and W. K. Hamblin, eds., Recognition of Ancient Sedimentary Environments, pp. 226-86. Society of Economic Paleontologists and Mineralogists, Special Publication 16.

Lowenstam, H. A. 1967. Adaptive Traits in Skeletal Morphology. American Geological Institute, Short Course Lecture Notes, Paleoecology, pp. HL1-HL13.

Schopf, T. J. M. 1969. "Paleoecology of Ectoprocts (Bryozoans)." Journal of Paleontology 43:34-144.

Valentine, J. W. 1973. Evolutionary Paleoecology of the Marine Biosphere, pp. 147-225. Englewood Cliffs, N.J.: Prentice-Hall.

Chapter Six

Paleoecology
Fossil Communities

Objectives

1. To become acquainted with some of the major types of benthic marine communities preserved in the rock record.
2. To learn what causes communities to change in both an ecologic and a geologic time frame.
3. To learn what new paleoenvironmental information can be gleaned from the analysis of ancient communities.

Important Terms

Community—A group of ancient organisms that lived together in a specific environment and interacted ecologically.
Diversity—The number of species which make up a community. Diversity varies among environments.
Ecosystem—The structure and ecological dynamics of a community, represented as the pathways of energy flow through the community.
Niche—The function of an organism within a community, especially its place in the food chain.
Resources—The supply of food, oxygen, and dissolved minerals needed by an organic community.
Succession—A continuous and deterministic series of changes in the structure of a community.
Trophic web—The food chain relationships within a community.

Introduction

In the previous chapters you have considered only the relationship between an individual organism and its physical environment. In this chapter you will realize that other organisms of the same and different kinds are also an important part of an animal's environment and that its relationships with other organisms provide new sources of ecologic information. The ecologic hierarchy (Table 6.1) is arranged in a similar way to the taxonomic hierarchy covered in Chapter 3.

TABLE 6.1. The Ecologic Hierarchy

Ecologic Group	*Ecologic Dynamics*	*Association*
Species (population)	Niche	Individual organisms
Community	Ecosystem	Populations
Province	Provincial system	Communities within a large geographic area
Planetary biota	Biosphere	All provinces

One of the most important aspects of a community is its diversity, which is a general guide to the complexity of the ecosystem represented. A second aspect to consider is the community's stability, that is, its tendency to keep the same association of species through time. Communities with high diversity are usually very stable. Communities that change their makeup in a regular and consistent way are said to undergo succession. The initial stage of succession is an unstable *pioneer* community, which eventually gives way to a final stable *climax* community. Some communities are so stable to begin with that they never experience succession. Stability depends not only upon the types of organisms present and their interactions, but also on the predictability of conditions in the physical environment.

The trophic base of any community is the assemblage of green plants (or plant plankton) that provides food—immediately to herbivores and ultimately to everyone else. Communities living near a river mouth may be subject to seasonal variation in the supply of dissolved nutrients (for example, phosphates and nitrates dissolved in river water), and they may be less stable than communities offshore that have less food but a more even supply. Unstable communities tend to have more individuals in each species, while stable communities have more species with fewer individuals in each. Diversity thus reflects the degree to which the available resources are partitioned among the niches; complex partitioning requires a stable, rather than a large, food supply. Conditions in the tropics tend to be less seasonal than in the temperate or lower latitudes, and the partitioning of resources is normally more intense. Polar ecosystems are seasonal in the extreme and support communities of very low diversity.

Detrital Substrate Communities

You probably have already seen fossils from communities that lived on muddy, silty, or sandy seafloors. Because they lived near sources of detrital sediment, they usually were subject to seasonal fluctuation in food supply and were generally less stable and less diverse than communities living far from any source of detrital sediment. A limy or carbonate substrate is basically formed by the skeletal material of organisms, and it usually indicates a community with a smaller, more stable food supply and high diversity. In some offshore areas the slow buildup of limy beds by highly diverse communities may alternate with relatively fast influxes of detrital sediment, causing local repopulation of seafloors by the more opportunistic organisms of unstable, low-diversity communities that normally live closer to the shoreline.

Because of the unpredictability of living in a detrital depositional system, the organisms in these communities are subject to stress from turbidity, burial by sediment, changing salinity and food supply, and variation in the availability of hard surfaces that sessile epifauna can attach themselves to. Because of continued stress and unpredictability over geologic time, the members of these communities rarely developed any specialized adaptations; they remained generalized animals that survived by reproducing prolifically but otherwise maintaining a semimarginal existence. Even though these species underwent normal periodic extinctions, their replacements in the same niche were commonly good carbon copies, or examples of evolutionary *convergence.* With a few important exceptions, the animals living in detrital communities today very much resemble those of the early Paleozoic era, both in morphology and in ecological functions.

Because of the general similarity of detrital communities throughout time, a few examples will suffice to illustrate the basic types. Examples of three basic detrital communities that lived in the Ordovician period are shown in Figure 6.1. The onshore community, which lived on sandy bottoms, was made up mostly of infaunal phosphatic brachiopods (lingulids), infaunal and epifaunal pelecypods, and epifaunal gastropods. Notice the large number of burrows in the sediment and the absence of a filter-feeding epifauna. An "intermediate" community was found either slightly offshore from the onshore fauna or along the shoreline away from river mouths. The silty substrate included a few infaunal pelecypods, scavenging gastropods and trilobites, and a filter-feeding epifauna of large flat brachiopods, clusters of small, ribbed brachiopods, a few bryozoan colonies, and crinoids. The offshore community, which grew on a muddy bottom, was the most stable and diverse assemblage, with few or no infauna, or scavengers, but a diverse collection of epifaunal filterers, including numerous bryozoans, brachiopods, and crinoids.

The general community types, illustrated by Ordovician examples, persisted throughout the Paleozoic era, at the end of which some important changes occurred. The trilobites and most of the bryozoans, brachiopods, and crinoids became extinct. Their vacant niches were filled in the Mesozoic era mostly by echinoids and suspension-feeding molluscs, which have persisted to the present day. Modern onshore communities look very much like those of the Ordovician period. The spatula-shaped phosphatic brachiopod *Lingula* is living today and has a fossil record that extends over 500 million years. The more stable offshore communities, paradoxically, have experienced the greater evolutionary turnover.

Carbonate Substrate Communities

The communities growing on limy seafloors present quite a contrast to the detrital communities because of the greater diversity of organisms, the different kinds of organisms present, and their strikingly more specialized adaptations. The discussion of these communities anticipates the following chapter, which deals with the origin of carbonate rock types. There is an intimate relationship between most carbonate sediments and the types of organisms growing on or near them. Carbonate substrate communities are characterized by a very stable suspended food supply and an overwhelming predominance of suspension feeders. These organisms have developed large massive skeletons that enable them to feed at many different levels above the substrate, thus achieving an extremely well developed partitioning of resources and indicating great trophic stability. Accumulations of these massive skeletons eventually build up as organic *reefs,* and skeletal disintegration produces carbonate sediments.

Like the detrital communities, many aspects of reef communities have not changed since they first appeared in the early Paleozoic era. Unlike the detrital communities, the reef assemblages show clearly developed stages of ecologic succession (Figure 6.2). The cause of reef succession is the growth of massive carbonate skeletons. Initially, these animals blanket an area of the seafloor; continued growth of the community produces a low moundlike bank that becomes more and more elevated above the seafloor (Figure 6.3A). As the mound grows higher, it becomes possible to exploit higher and richer levels of suspended food in the water, thus increasing the number of niches and achieving a greater partitioning of resources (Figure 6.3B). The final stage in reef growth occurs when the mound reaches the level of low tide and cannot grow higher and maximum resource partitioning has been achieved (Figure 6.3C). Notice in these assemblages the highly specialized adaptations for filtering suspended food from the water.

In the Silurian reef communities illustrated, the pioneer stage consists of filter-feeding tabulates, bryozoans, brachiopods, crinoids, and a few scavenging trilobites and snails. All of these

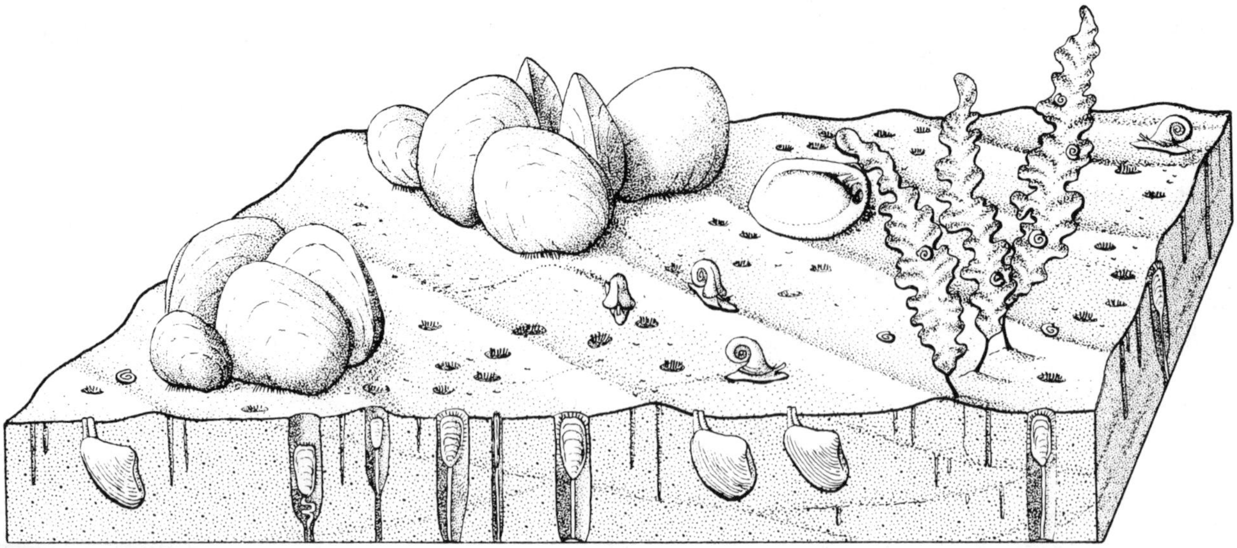

Onshore community

Intermediate community

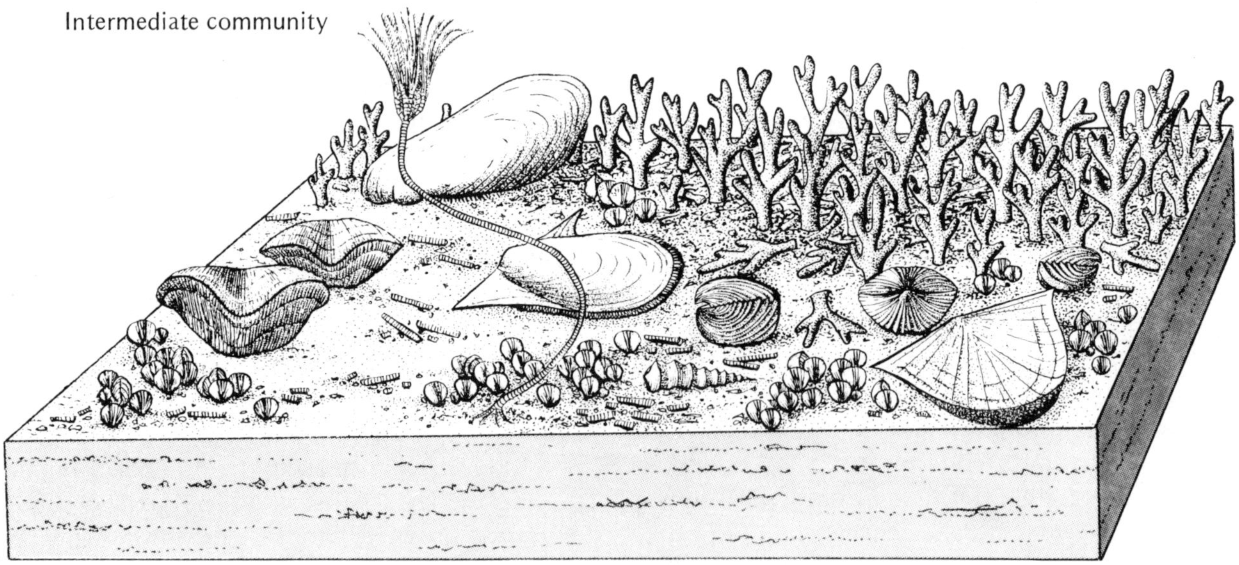

Offshore community

Figure 6.1. Reconstructions of major types of detrital substrate communities in Late Ordovician rocks of the Appalachian area.

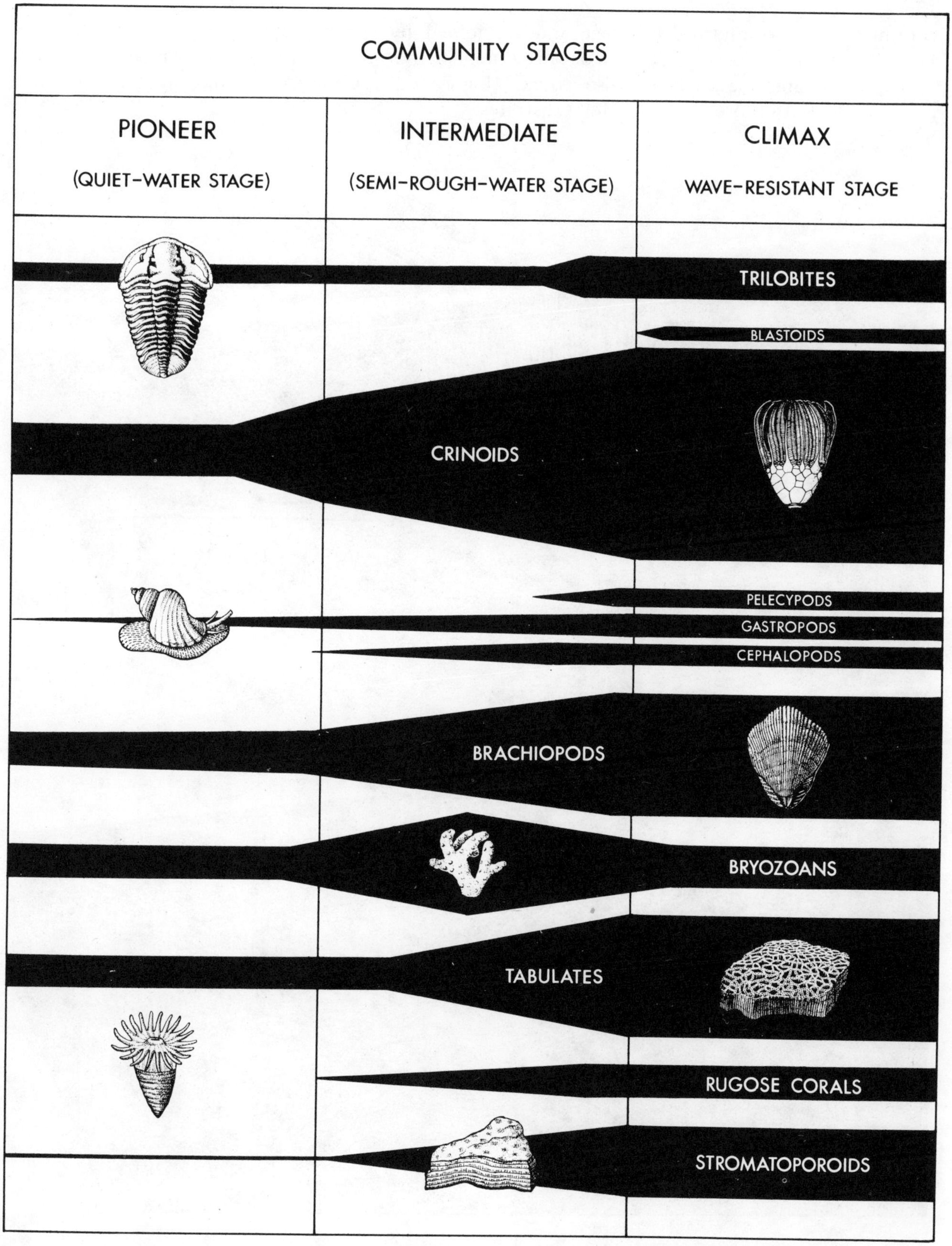

Figure 6.2. Succession of community types in Silurian organic reefs, showing changing proportions of animal groups and diversity increasing to the right.

continue into the intermediate stage and are joined by stromatoporoids and the most diverse assemblages of bryozoans. The bryozoans diminish in the climax stage, in which pelecypods, cephalopods, and blastoids are also found (Figure 6.2). Contrast the feeding types in these communities with those on the detrital substrates.

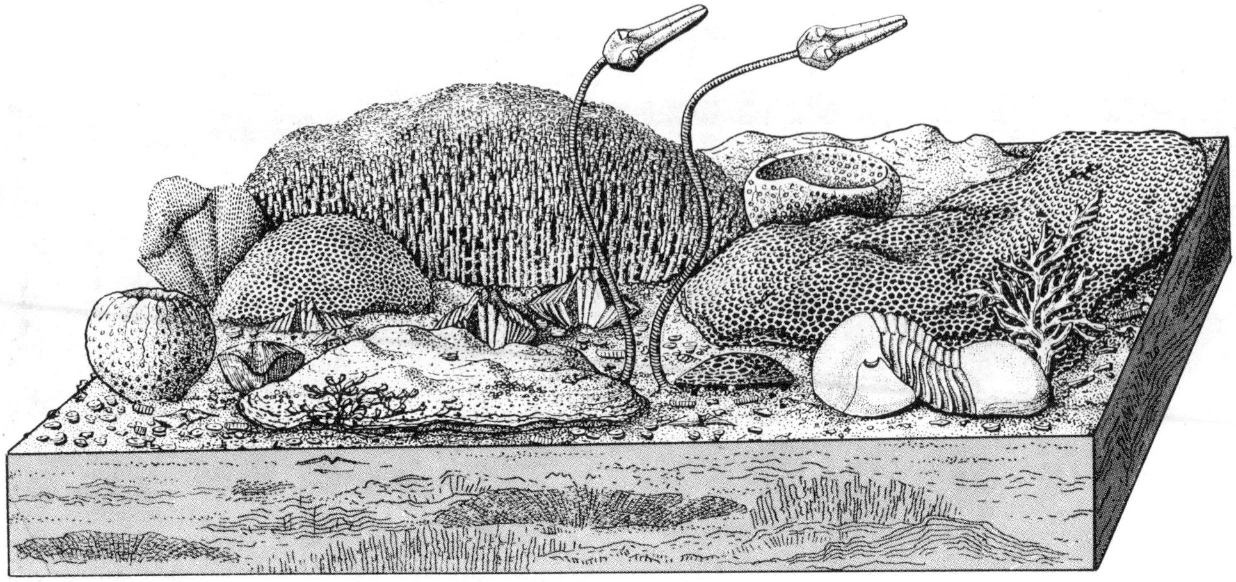

A. Pioneer stage

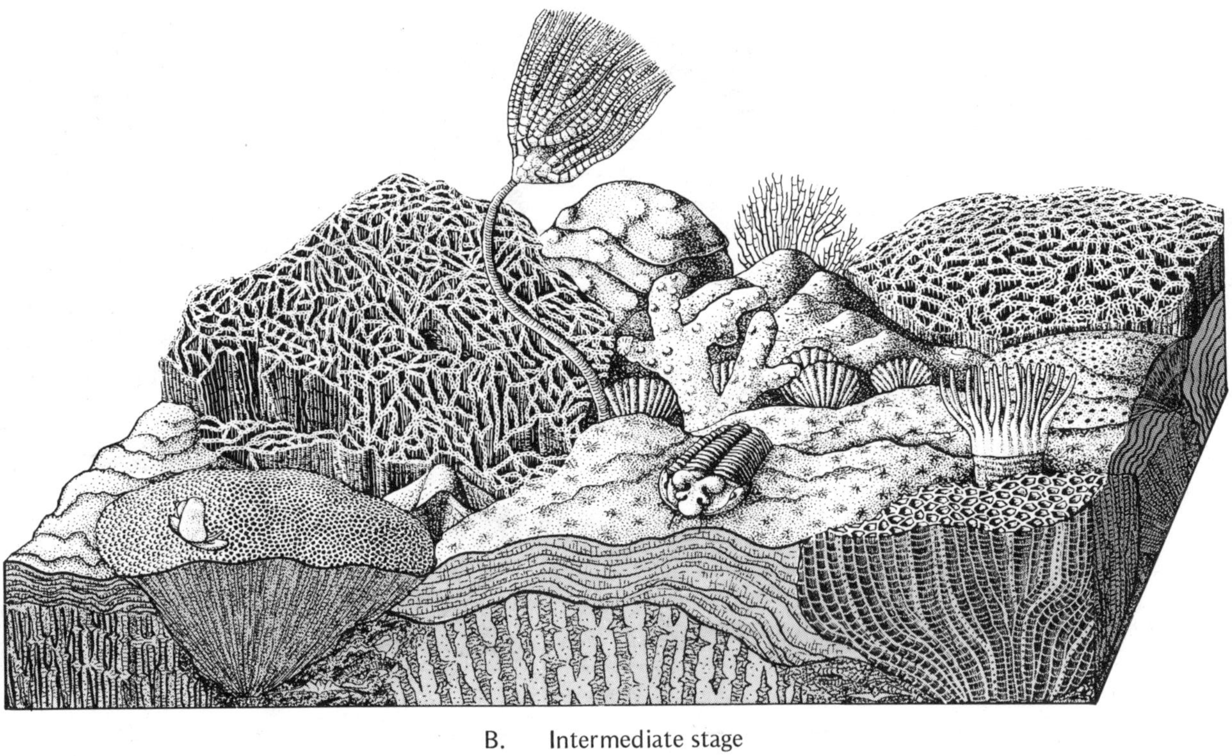

B. Intermediate stage

Figure 6.3. Succession of middle Silurian organic reef communities in the Great Lakes region of North America.

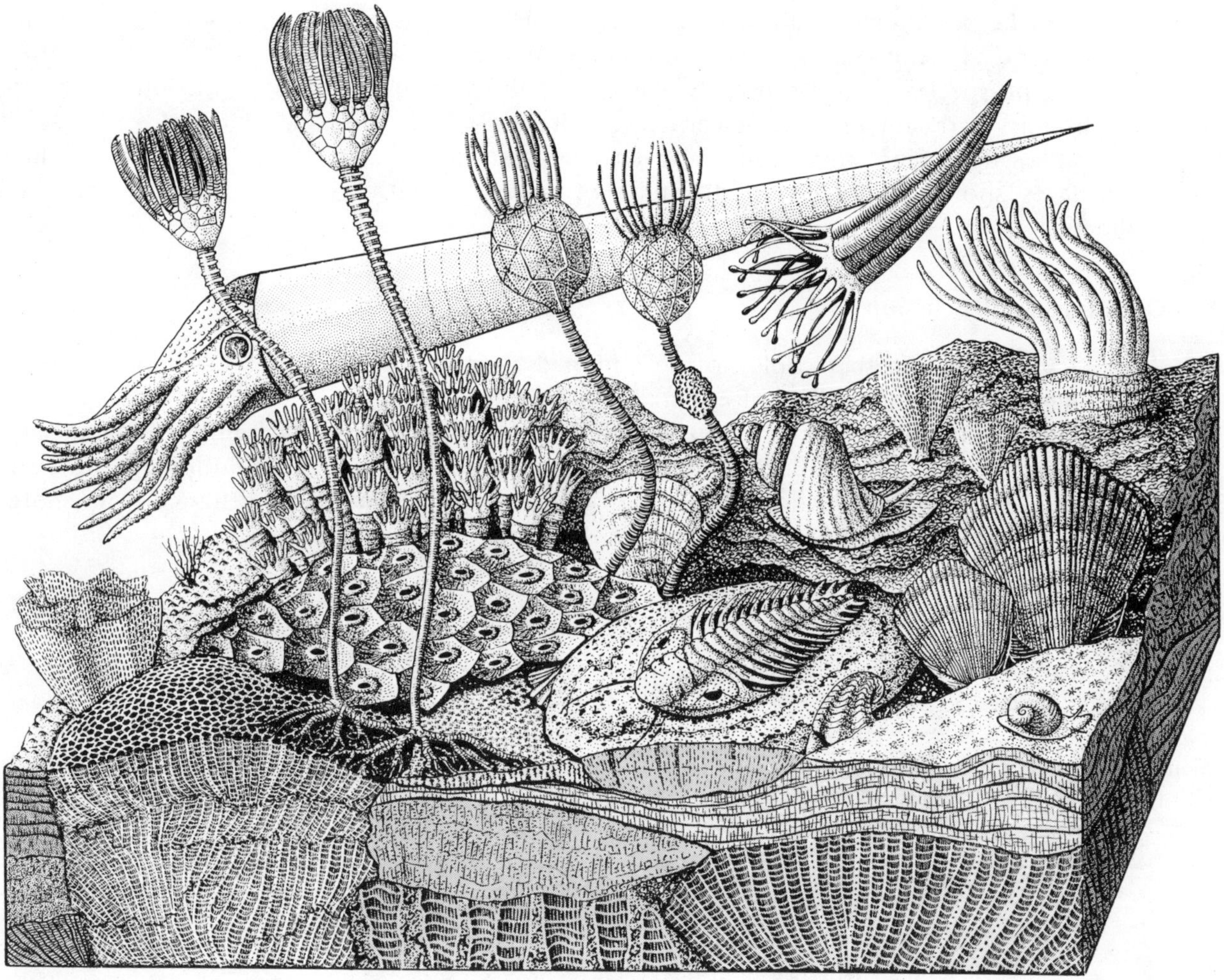

C. Climax stage

Figure 6.3 continued

Throughout geologic time, reefs have varied in their dominant framework-building organism. The Silurian reefs shown were dominated by stromatoporoids and tabulates. In other Paleozoic reefs, the dominant building organisms were calcareous algae or bryozoans. Like the detrital communities, these general reef types lasted through the Paleozoic era but underwent some major changes afterwards. The tabulates, stromatoporoids, and bryozoans were all involved in the mass extinction at the close of the Permian period. In the Mesozoic era, the scleractinean corals appeared and became the most important reef builders. They had the unique feature of symbiotic algae imbedded in their tissues, which helped them produce more massive and larger reefs than any that lived in the Paleozoic era. The scleractinean reefs living today contain many cavities and crevices, and in some of these cryptic habitats are found assemblages of sclerosponges, brachiopods, and bryozoans. Thus the Paleozoic reef communities may still survive in crevices in living reefs.

Exercise

A collection of slabs containing natural assemblages of detrital and carbonate substrate fossils will be set out in the laboratory (ideally, they will be different from those used in the preceding chapters).

1. Determine the percentage of all feeding types in each slab, as well as the rock type. Identify the community type or stage of succession represented by each.
2. Calculate the species diversity in each slab, and rank the slabs according to trophic stability. Does diversity correlate with the percentage of filterers?
3. If possible, determine the height above the substrate of each type of filterer. Which assemblages show the greatest partitioning of resources with respect to height above the sediment?

Questions for Discussion

1. What new information that could not be obtained by studying individual fossils or rock types can be derived from the analysis of communities?
2. Will evolution take place at a faster rate in trophically stable or unstable communities? What is the relationship between the level of environmental stress and rate of evolutionary turnover?
3. Can you use community structure to identify water depth, climate, or distance from shore? What are the limitations on interpretations of this kind?

Background Reading

Bretsky, P. W. 1969. "Evolution of Paleozoic Benthic Marine Invertebrate Communities." Palaeogeography, Palaeoclimatology, Palaeoecology 6:45-59.

Newell, N. D. 1972. "The Evolution of Reefs." Scientific American 227:54-65.

Valentine, J. W. 1973. Evolutionary Paleoecology of the Marine Biosphere, pp. 269-336. Englewood Cliffs, N.J.: Prentice-Hall.

Chapter Seven

Sedimentary Petrology Origin of Carbonate Rocks

Objectives

1. To understand the origin of the grains, matrix, and overall textures of carbonate rocks.
2. To relate the textural characteristics of the different rock types to environmental factors such as water movements affecting the sediment, plant growth, other biotic activities, and seawater chemistry.

Important Terms

Allochems—The grains of a carbonate rock, analogous to sand grains or pebbles in a detrital rock. Allochems include fossils, oolites, intraclasts, and pellets.

Aragonite—Mineral calcium carbonate, $CaCO_3$, having an otthorhombic crystal lattice. Aragonite usually recrystallizes as calcite over long periods of geologic time.

Calcite—Mineral calcium carbonate, $CaCO_3$, having a rhombohedral crystal lattice.

Dolomite—Mineral calcium magnesium carbonate, $CaMg\,(CO_3)_2$.

Dolostone—Rock composed primarily of dolomite.

Intraclasts—Eroded lithified fragments of a preexisting rock or sediment. Examples include the broken-up pieces of formerly flat beds in a "flat pebble conglomerate" (see Figure 8.7) and reworked clusters of lithified oolites or pellets.

Limestone—A rock type arbitrarily defined as composed of more than 50 percent aragonite and/or calcite by weight.

Matrix—Either fine-grained material between allochems (usually *micrite* or lithified lime mud) or crystalline calcite cement (*spar)* filling spaces between grains.

Oolites—Small spheres the size of coarse sand. In cross section they are made up of either concentric layers or radiating fiberlike crystals of calcium carbonate. They are formed by direct precipitation of calcium carbonate from seawater.

Pellets—Round grains the size of coarse sand, composed of homogeneous lime mud, usually the carbonate fecal material of sediment-ingesting animals.

Introduction

Carbonate rocks are formed from sediment that originates at or very near its place of deposition. Detrital rocks, in contrast, are composed of materials that originated at some considerable distance away and were carried to the depositional site. Detrital rocks provide information about (1) the source area, (2) transport processes, and (3) the depositional environment. In carbonates all three of these are one and the same.

Limestones are composed of calcite and aragonite. Calcite is much more common, because aragonite is unstable and slowly recrystallizes to calcite over time. No Paleozoic limestones contain any aragonite, and very few Mesozoic ones do. The mother-of-pearl layers that you see on some mollusc shells are aragonite. Dolomite is formed in certain environments in which calcium carbonate reacts with the magnesium in seawater (see Chapter 11). You can recognize some dolostones easily because they look sugary or granular, often contain many small cavities, and any fossil fragments in them are distorted by recrystallization. Others look just like many limestones and must be identified by chemical tests.

One of the most important things to realize about carbonates is that most of them would never have formed without the abundant growth of plants and animals in the depositional environment. Almost all of the Precambrian carbonates, those formed earlier than 600 million years ago, are known to be closely tied to the activity of blue-green algae (see Chapter 10). Phanerozoic limestones include a new type, *skeletal* limestone, formed from the disintegration of the skeletons of invertebrates and higher algae. Pelletal limestone is also a uniquely Phanerozoic rock type. Oolitic and intraclastic limestones and dolostones are found throughout the rock record (Figure 7.1).

Another important concept is that carbonates form in almost all aquatic depositional environments and are originally incorporated into almost all sedimentary rocks. However, when streams or currents supply detritus, the carbonate material is correspondingly diluted and a mixed carbonate-detrital rock is formed (Figure 7.2). The *absence* of detritus is required to produce a pure carbonate, which can only accumulate along coastlines or on marine shelf areas far removed from a river mouth or longshore current. Some of the only places in North America where pure carbonates are being deposited today are the Florida Keys and the Bahamas. By looking at a map (see Figure 17.5) you can see that both areas are well removed from the supply lines of detrital sediment.

A stamp collection can be arranged in many different ways: by color, subject, shape, country, or year. Classifying carbonate rocks is a similar problem in that there is no overall method that is best for every purpose. The classification method used in this chapter has been selected for its simplicity and is based upon four types of grains imbedded in two types of matrix material (Figure 7.1). The four grain types are fossil fragments, oolites, intraclasts, and pellets; and the matrix is either lime mud, which lithifies to micrite, or clear, sparry calcite cement (spar). Lime mud resembles detrital mud or silt in appearance. A simple identification method is to test a cleanly broken surface with a tiny drop of dilute hydrochloric acid: micrite and spar will effervesce vigorously, but detrital silt or dolomite will not. Study the acid reaction under a microscope to be sure that the matrix itself, and not just the nearby grains, is effervescing. Spar will show, in a freshly broken surface, the shiny cleavage faces of small calcite crystals. Be careful not to confuse the calcite cleavages in small fossil fragments, such as crinoid stems, with sparry cement. You can see that the microscope is a necessity in identifying limestones.

This method of classification assumes that most carbonates are *bimodal;* that is, they have both fine and coarse material, without intermediate sizes, in the same sediment. This is especially true of skeletal limestones, because organic skeletons break down into bimodal sediments consisting of large skeletal sections or pieces (allochems) and a grit or dust (matrix) that formed as the larger pieces broke apart (Figure 7.3). Green algae (Figure 7.4A) are a main source of lime mud in many modern carbonate environments. Algae and other plants (Figure 7.4) tend to stabilize the fine-grained lime mud even in areas with much water movement, making it difficult to use the grain size of carbonate rocks as a guide to the depositional environment.

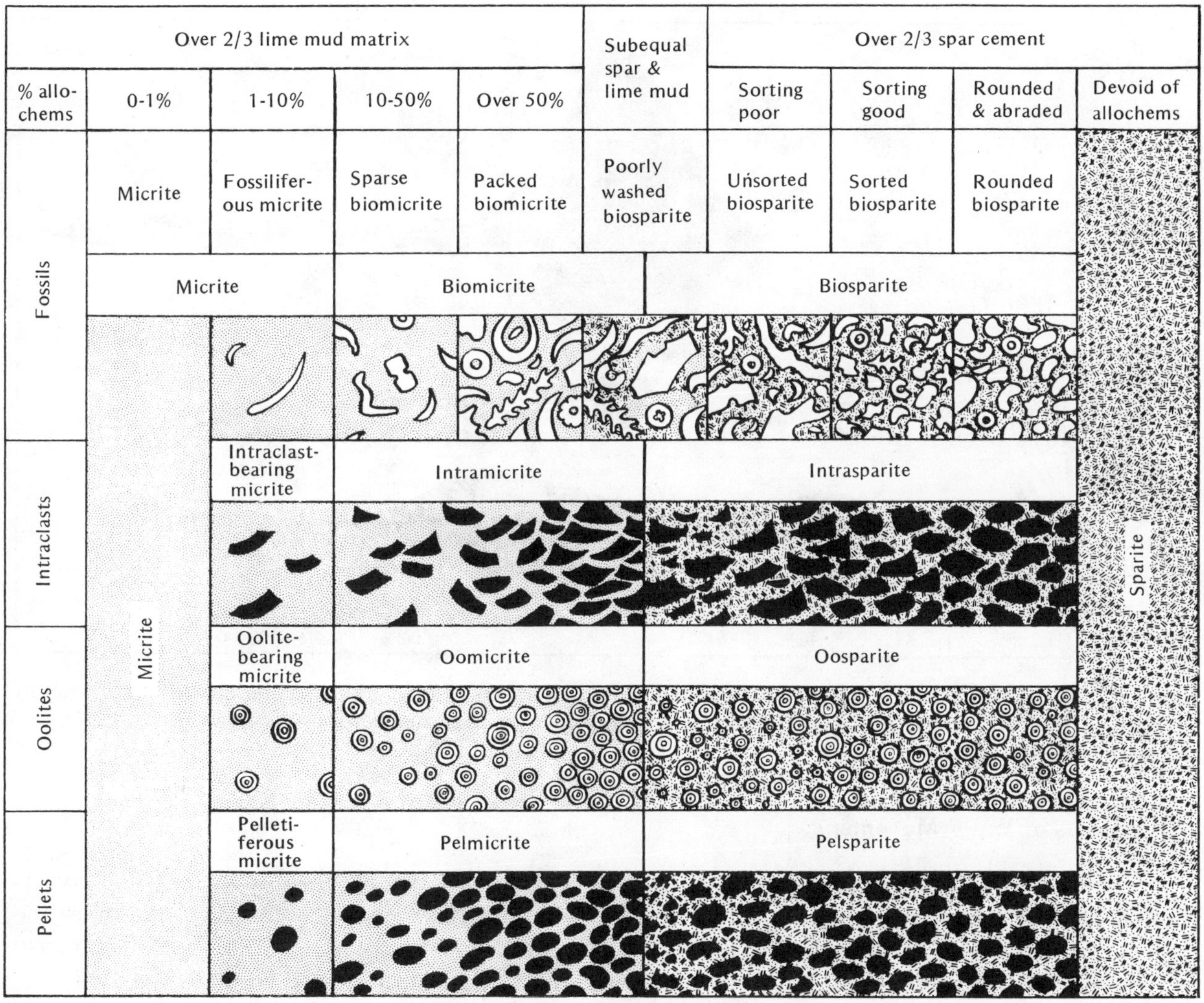

Figure 7.1. A classification of limestone types. Mixed types have names such as fossiliferous intrasparite, oolitic pelmicrite, pelletiferous oosparite, or intraclastic biomicrite.

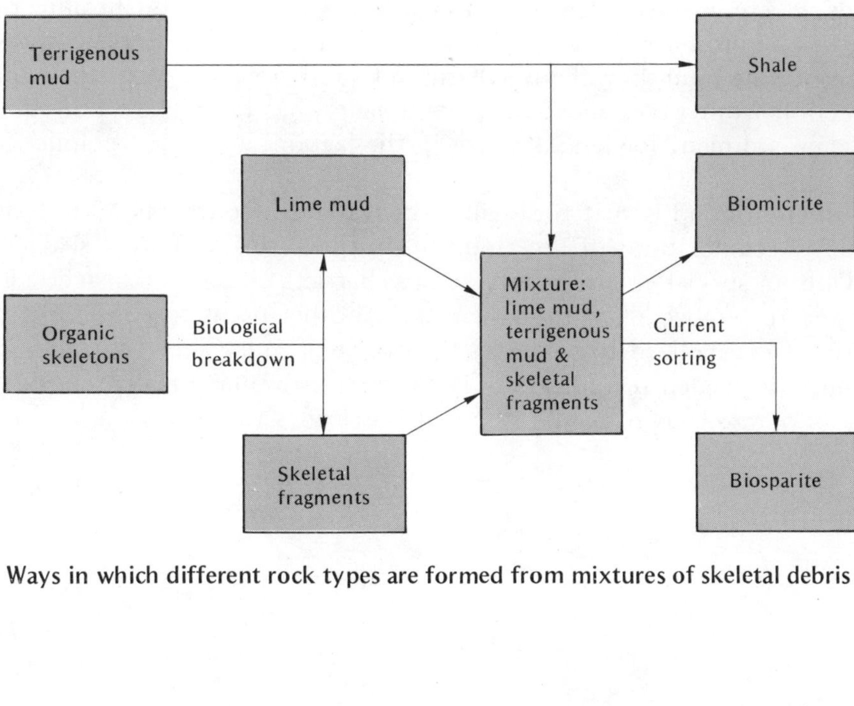

Figure 7.2. Ways in which different rock types are formed from mixtures of skeletal debris and detrital mud.

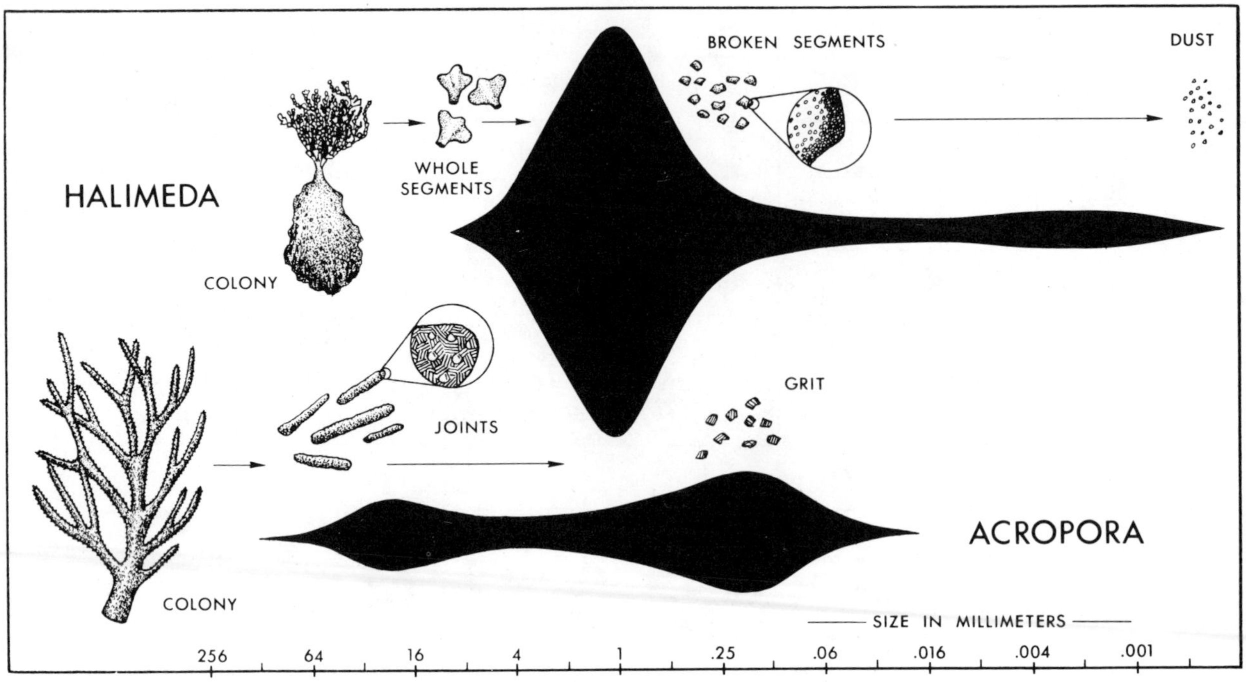

Figure 7.3. Many carbonate sediments have large allochems in a matrix of finer-grained material. Such sediments are the natural result of the way many organic skeletons disintegrate, as illustrated here for the green alga *Halimeda* and the colonial coral *Acropora.*

Effects of Water Movement

Spar is formed when void space in a coarsely grained carbonate sediment is filled by the growth of calcite cement after deposition. It generally indicates, especially in a skeletal limestone, that any original lime mud has been washed away by water movement. Thus rocks with less micrite and more spar probably represent environments with more water movement, which winnowed away some of the lime mud. Figure 7.5 shows a general profile of the types of water movement and the resulting carbonate rock types. A specific illustration, the carbonate sediments of the Great Bahama Bank, is given in Figure 7.6. An elongate oolite bar lines the western margin of the bank, the side of the prevailing winds, where water movement is the strongest. The continual rolling of grains in the carbonate-saturated water produces the concentrically laminated oolites. Fragmented skeletal debris and intraclasts, free of lime mud, line the northern and southern margins of the bank. The central lagoon, with protection from water movement on all sides, has the highest buildup of lime mud. Many pellet-producing sediment feeders also live in the lagoon, where the pelmicritic sediment is stabilized by turtle grass.

One point to keep in mind is that the shallowest water environments do not always have the most agitated water. Accumulations of lime mud in general probably reflect shallow water rather than deep water. Only in special environments, such as barriers or exposed beaches, is the sediment sufficiently winnowed to remove all of the lime mud. Another point to consider is that micrite in ancient rocks may have crystallized to spar over the course of time. Therefore, additional evidence, such as fossil evidence, is needed to indicate quiet or agitated water. Finally, consider what effects on sediment distribution are caused by the frequent hurricanes that cross the Great Bahama Bank every year.

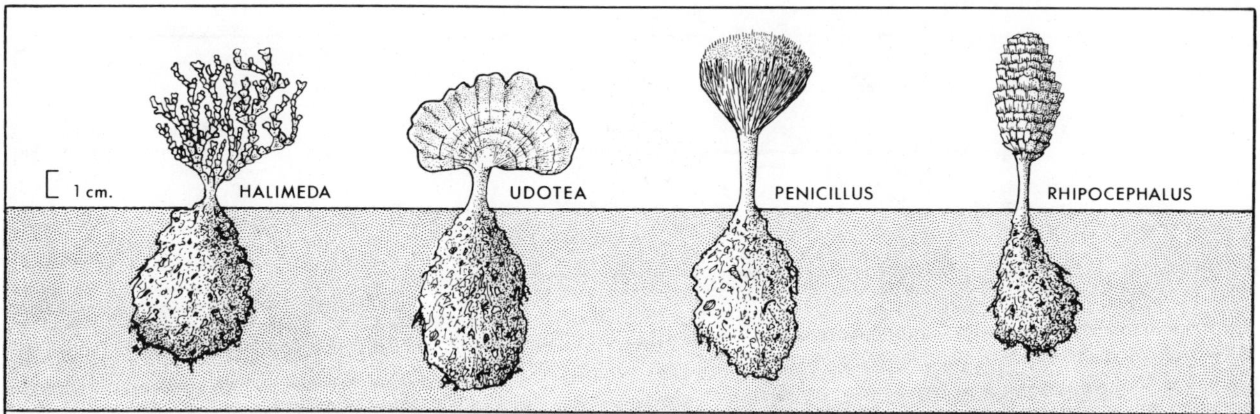

A. Types of calcareous algae which bind sediment by means of a buried holdfast; breakdown of the algal skeletons produces much of the lime mud that will lithify to micrite.

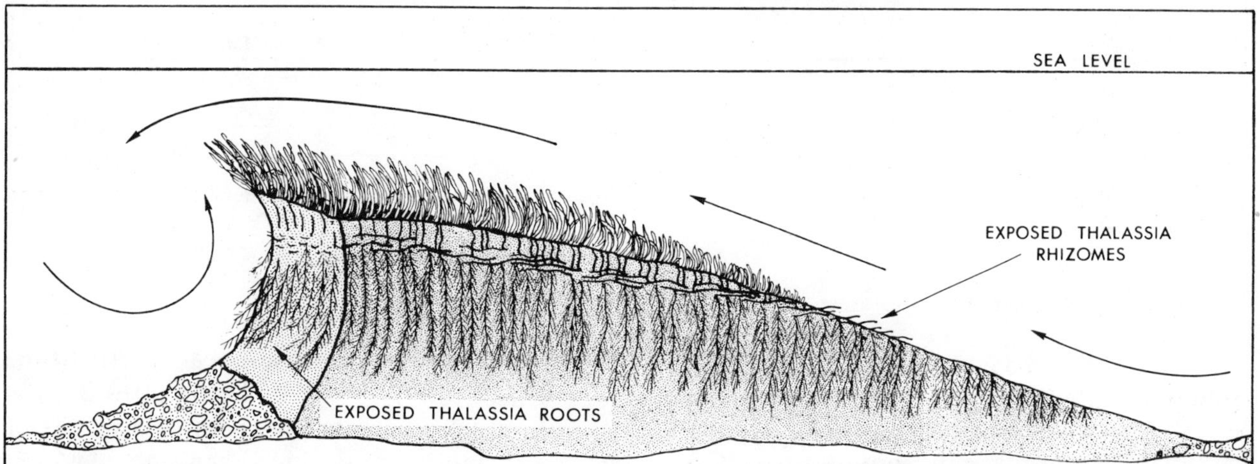

B. Mound of sediment bound by the roots and rhizomes of turtle grass *Thalassia testudinum.*

C. The thick tangle of red mangrove *(Rhizophora mangle)* roots plays an important role in the formation of new land by catching and holding sediment.

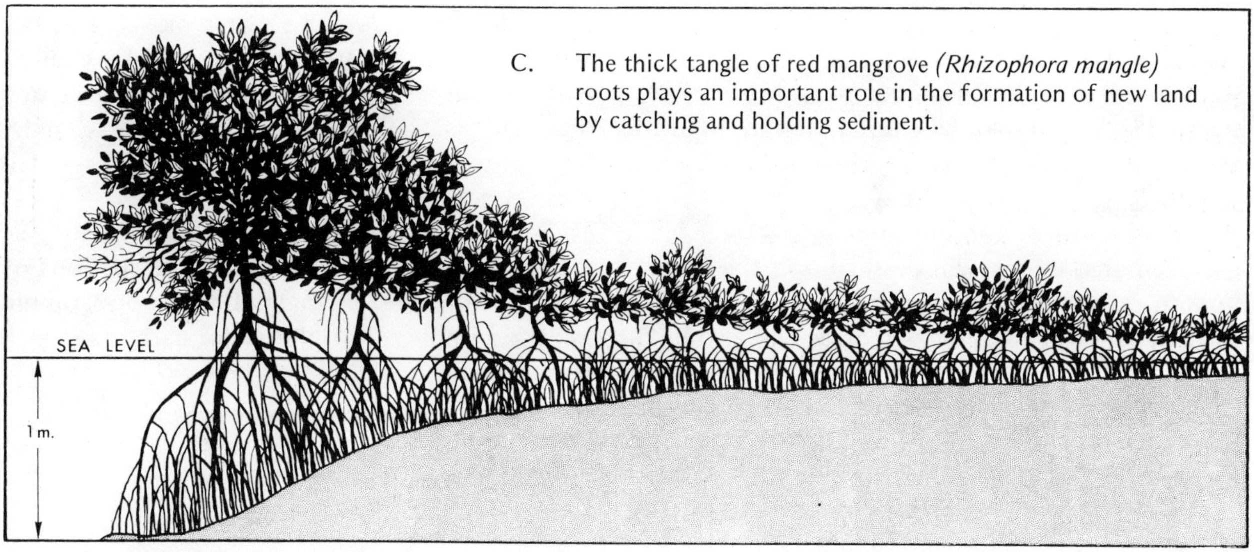

Figure 7.4. Plants that trap lime mud in modern carbonate environments in Florida and the West Indies.

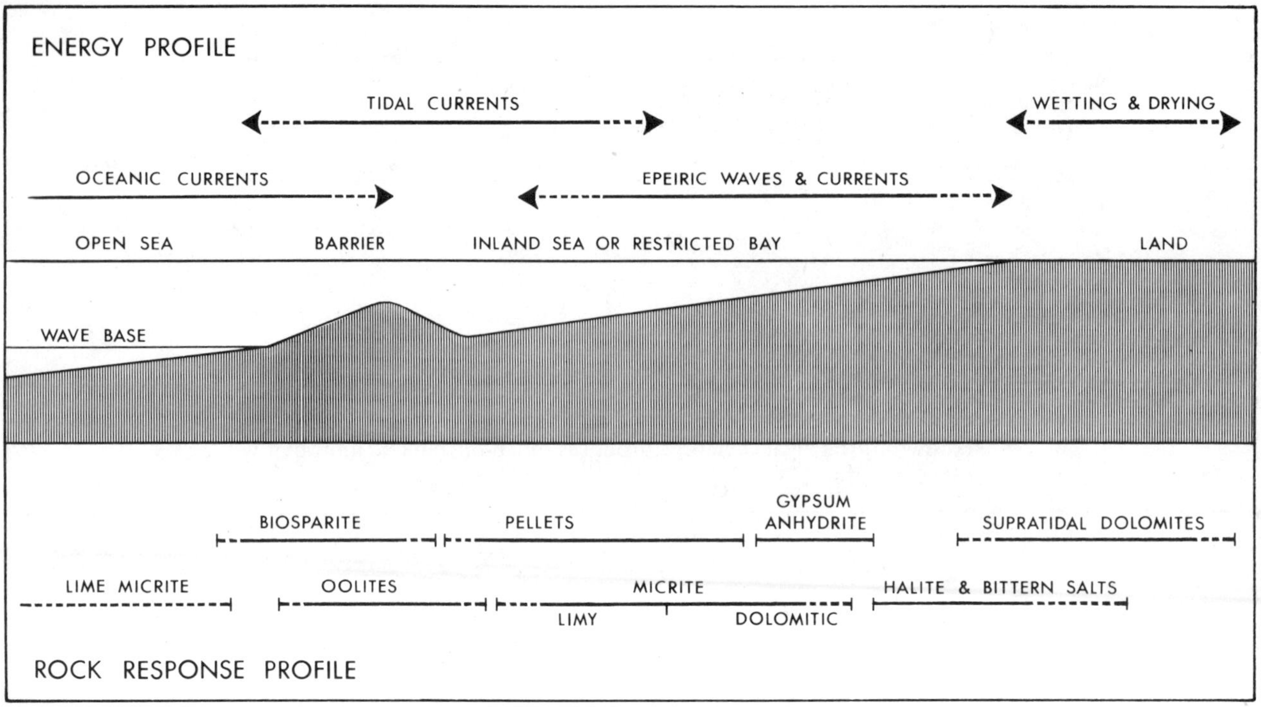

Figure 7.5. Simplified profile of water movements and carbonate sediments across a marine shelf area having no supply of land-derived detritus.

Effects of Organisms

Although the rounding and sorting of shell fragments in carbonates may appear similar to the rounding and sorting of detrital sediments, it is misleading to interpret carbonate textures only in terms of water movement. The organic communities not only produce the allochems and matrix of carbonate sediments, but continue to act on them after deposition. Boring organisms, including algae and sponges, can thoroughly perforate a calcareous shell, not only producing lime mud directly, but hastening the final disintegration of the skeleton into sediment. The size, sorting, and roundness of a carbonate sediment may depend upon the makeup of the organic community that produces it. Bryozoan and algal skeletons may disintegrate with very little water movement, but snail and clam shells are very durable and can survive considerable agitation with very little change. Sediment feeders like sea cucumbers may cause the sorting out of particles too large to ingest, and ingested sediment may be altered even more in their digestive tracts.

Exercise

Demonstrations will be set up in the laboratory to illustrate map views and cross sections of modern carbonate environments such as the Florida Keys and the Great Bahama Bank. Topographic and bathymetric maps of such areas will also be displayed. Examples of all the major carbonate rock types will be keyed into their places of origin in the demonstrations.

1. You will be given a set of unknown carbonate rocks. Using a binocular microscope (or hand lens), record the following for each unknown.
   a. All types of allochems present and their proportions.
   b. Relative proportions of spar and micrite.
   c. Rock name (from Figure 7.1).
   d. Fossil types identifiable from skeletal fragments.

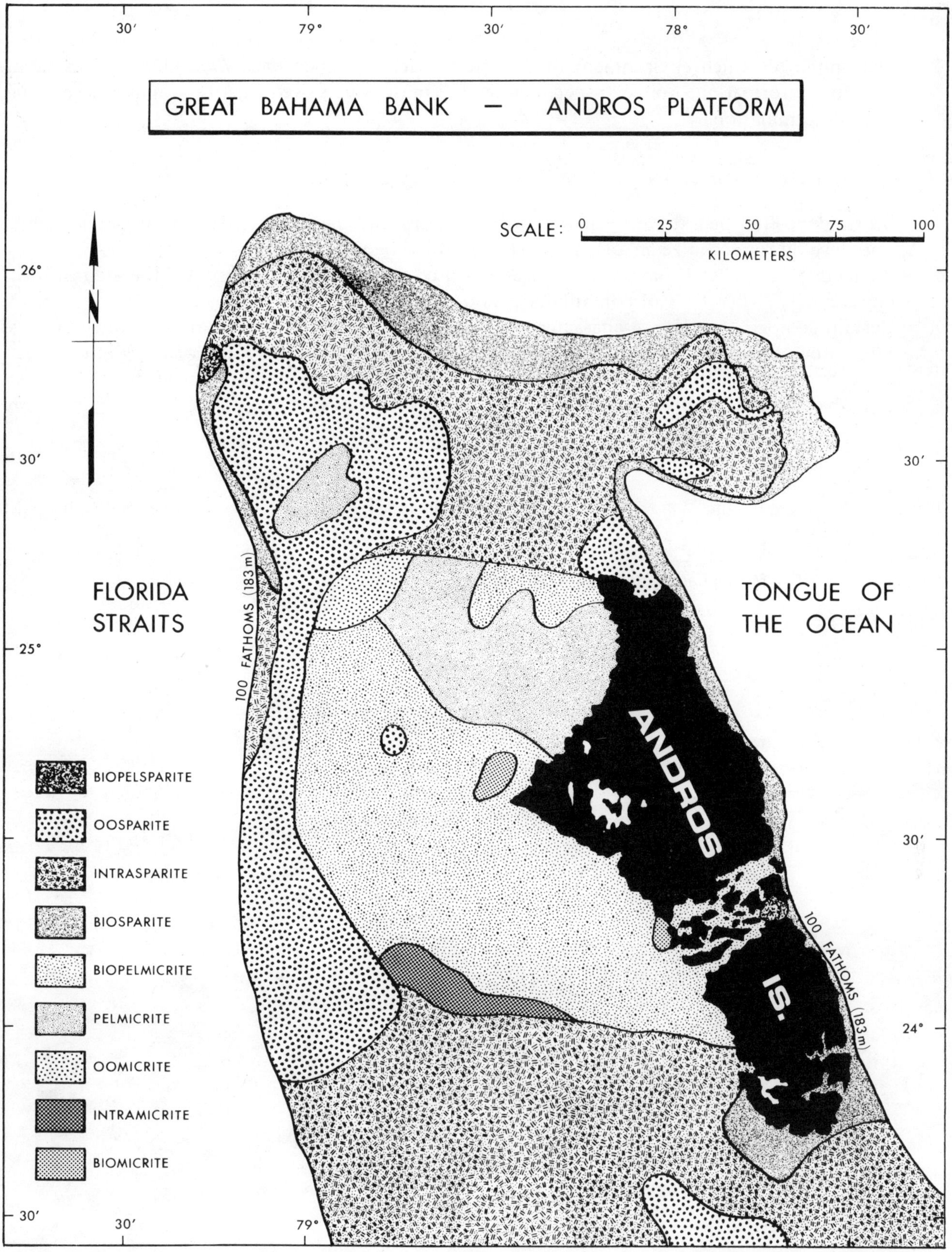

Figure 7.6. Map of contemporary carbonate sediments accumulating on the Great Bahama Bank. The only part of the bank above water is Andros Island. The areas shown as white are deeper water than the edge of the continental shelf.

2. Determine to which environment of deposition each specimen *most likely* belongs, using all available information. For each specimen list the most conclusive evidence you used in making your determination.

Questions for Discussion

1. Storm-deposited beds have been mentioned in this and earlier chapters. What criteria could you develop to recognize an ancient storm-deposited bed?
2. What does a pelletal limestone indicate about the trophic stability of the depositional environment? What type of community produced it?
3. Certain geologic periods are characterized by abundant carbonate rocks and very few detritals. What does this indicate about the paleogeography of those time intervals? What does it indicate about the trophic stability of the biosphere? Can you suggest reasons for this related to global climate and plate tectonics?

Background Reading

Blatt, H., Middleton, G., and Murray, R. 1972. Origin of Sedimentary Rocks, pp. 409-76. Englewood Cliffs, N.J.: Prentice-Hall.

Horowitz, A. S., and Potter, P. E. 1971. Introductory Petrography of Fossils, p. 2-33. New York: Springer-Verlag.

Milliman, J. D. 1973. Marine Carbonates. New York: Springer-Verlag.

Chapter Eight

Sedimentology Inorganic Sedimentary Structures

Objectives

1. To recognize the shape and form of grain aggregates known as *sedimentary structures.*
2. To understand the physical processes that produced such structures.
3. To determine their usefulness in the interpretation of ancient depositional environments.

Important Terms

Bedding—The layering of sedimentary aggregates. Layers (*strata)* range in thickness from a few millimeters to several meters. Very thin strata in the range of a few millimeters are called *laminations;* thicker units, *beds;* and discontinuous beds, *lenses* or *lentils.* Cyclic alternation between fine and coarse-grained laminations are *varves* (Figure 8.1).

Cross bedding—Bedding that is not horizontal, caused by deposition on a sloping surface, as on a ripple ridge in a stream or on the sloping face of a sand dune.

Desiccation cracks (mudcracks)—Vertical V-shaped cracks caused by the drying out of a sediment layer, separating it into a network of polygonal plates that may be broken up by later water movement and redeposited as intraclasts in a *flat* pebble conglomerate. Sometimes water movements stack these plates up against one another like shingles, producing an "edgewise" conglomerate.

Raindrop prints—Along with mudcracks, raindrop and hailstone imprints reflect the exposure of wet sediment to the air before it hardened and became buried. Such conditions occur in very few depositional environments.

Ripple marks—Parallel ridges of sediment that are usually perpendicular to the movement of water or wind. Changing current directions may superimpose a second set of ripples on a previous one, producing rhomb-shaped interference ripple marks.

Salt casts—Cubic cavities within a sediment that were once occupied by salt crystals.

Sole marks—Parallel ridges on the underside of a bed that are usually parallel to the direction of water movement.

Introduction

Sedimentary structures result from the nature and intensity of the movement of sediment-laden water or air, or from such simple physical processes as the drying out of wet mud, or raindrops falling on a sediment surface. Fluid flow can be very even, like honey flowing from a jar, or very choppy and irregular, like curls and billows of smoke emerging from a stack. If the fluid is actively depositing sediment, then the piles or layers of sediment that it leaves will reflect the evenness or the turbulence of the flow. The form of the deposited sediment reflects not only the velocity of the flow but also its patterns of regularity or irregularity. Figure 8.1 shows a variety of bedding features. Ripple and sole marks result from directional water movement in which turbulence or irregular flow patterns have developed in water moving over the bottom. The turbulence is developed as eddies or vortices (Figure 8.2); because of changing velocity in the turbulent eddies, sediment may be deposited between them, forming ripple marks (Figure 8.4). If

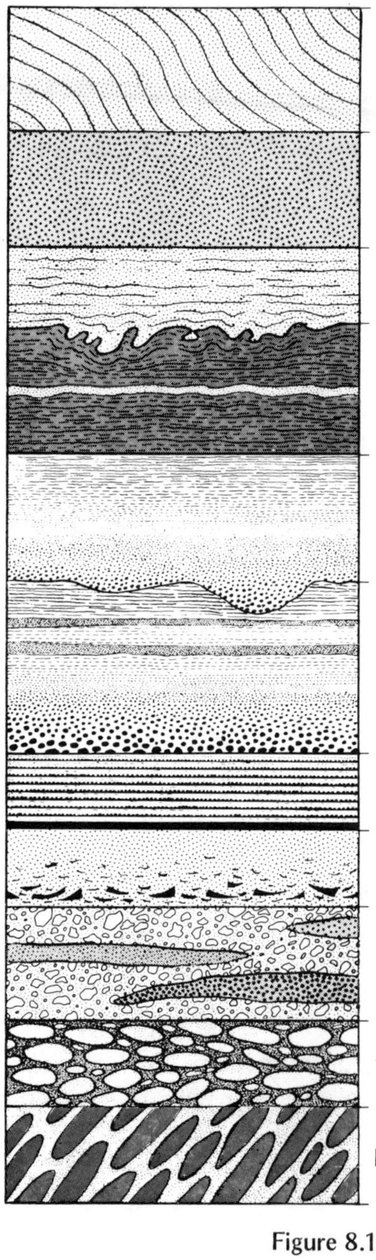

A. Sandstone with cross beds.

B. Massive bed of homogeneous sandstone.

C. Sandstone with horizontal bedding; post-depositional deformation structures on lower surface.

D. Shale with mica flakes parallel to bedding, and containing a thin sandstone bed.

E. Graded bedding, ranging from coarse sediment below to fine above; lower surface has been scoured and filled.

F. Graded bed with sandstone laminae appearing in the shale.

G. Varves, alternating fine- and coarse-grained laminae.

H. Sandstone bed containing flat pebbles of hardened mudstone or shale.

I. Conglomerate bed containing lenses of sandstone.

J. Conglomerate with pebbles roughly parallel to bedding surface.

K. Edgewise, or imbricated conglomerate.

Figure 8.1. Types of beds and bedding.

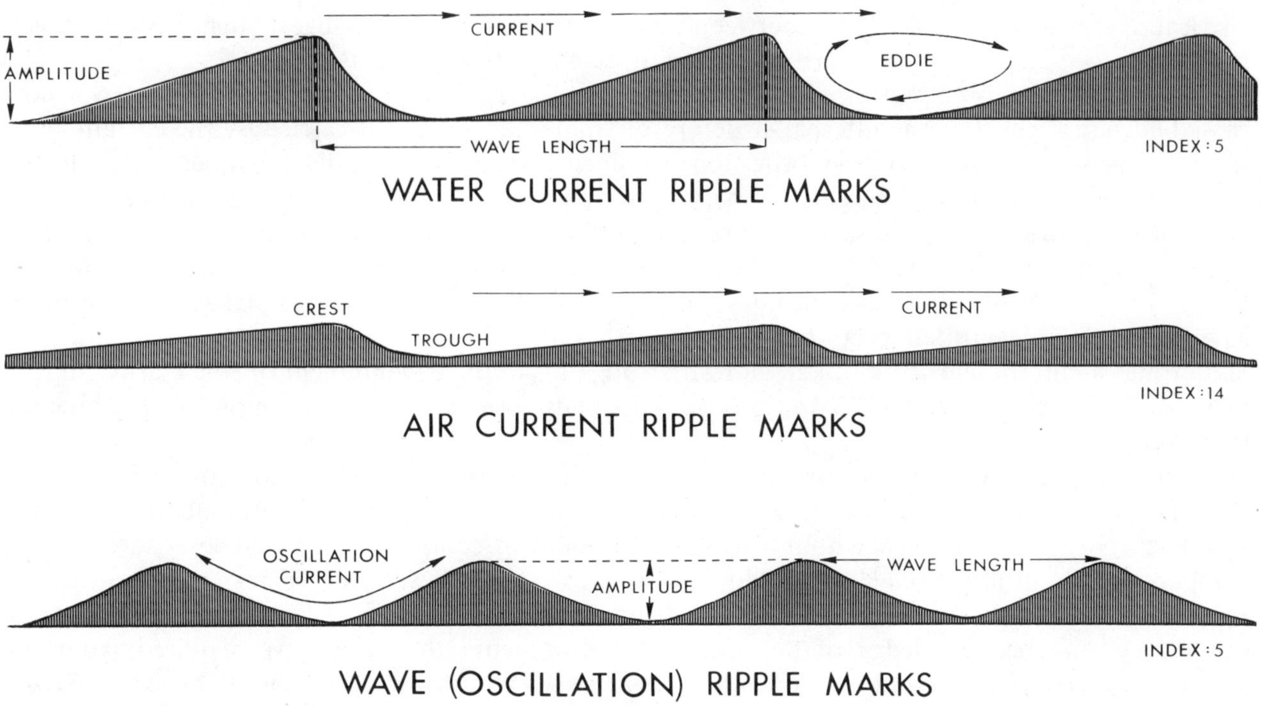

Figure 8.2. Cross sections of ripple marks. The index is the ratio between the wave length and the amplitude.

the vortices are moving parallel to the overall flow, they may scour out troughs in the bottom, which, when buried by an overlying bed, will show up in reverse relief as sole marks (Figure 8.3). You will observe that ripple marks produced by a current are not symmetrical. Wind-driven waves at sea, however, may produce oscillations in the water that will affect marine shelf sediments down to a depth of about 35 meters, a level known as *wave base.* Ripple marks caused by such oscillations are more symmetrical than those formed in a current. Vortices moving parallel to the current may

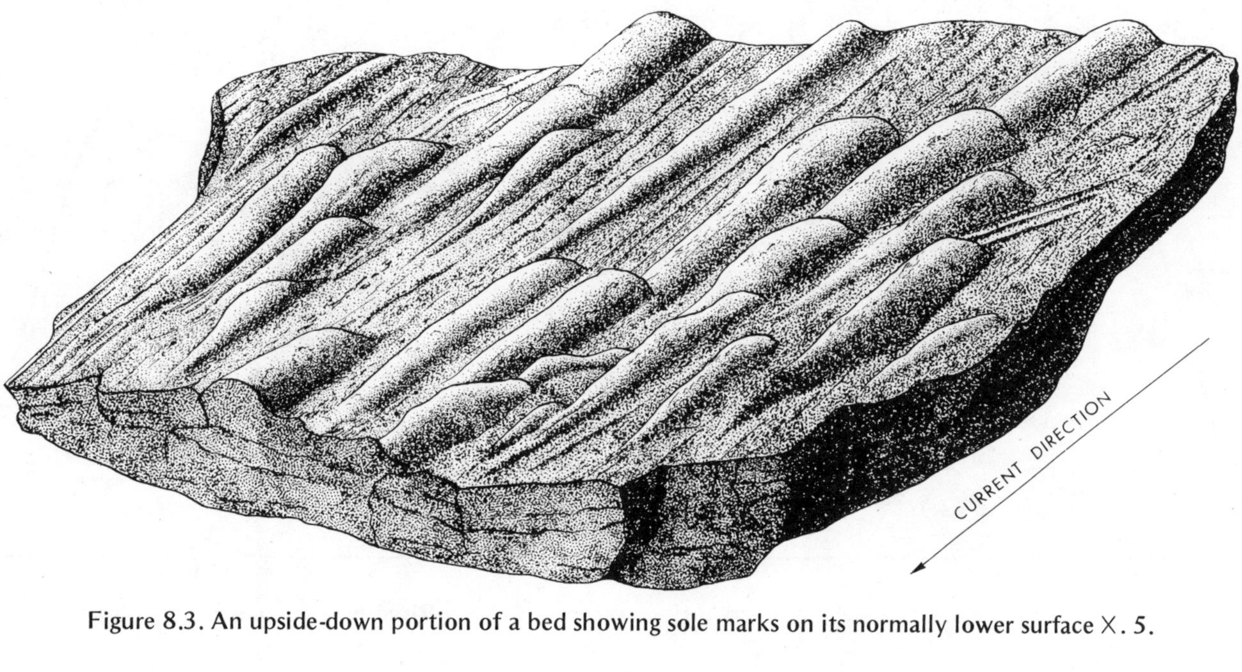

Figure 8.3. An upside-down portion of a bed showing sole marks on its normally lower surface X. 5.

also cause sediment to be deposited between them, in long symmetrical ridges. Such deposits, called *tidal sand ridges,* are common in shallow waters affected by tidal currents.

Cross bedding is caused by deposition on a sloping surface. You can see in Figure 8.4 how cross beds are formed as the internal structure of ripple ridges. Sand is carried by the current away from the gentle backslope and deposited on the steeper foreslope. Usually cross beds slope in the direction of current flow (Figure 8.6). An exception, however, occurs when ripple-forms are eroded on their downstream side and sand is deposited on the upstream side; such structures are *antidunes.* The cross bedding in fluvial point bars is usually at right angles to the overall flow of the stream (see Figure 14.3). As sedimentation continues, one set of ripples or dunes may be planed off by erosion as another set is deposited over them, resulting in superposed groups of cross beds (Figure 8.5). Sometimes a current will scour out an elongate trough (as in the development of sole marks) and fill it in with cross beds (Figures 8.5 and 8.6B). Cross beds formed by wind are more steeply inclined than those in water.

Other primary sedimentary structures include mudcracks, flat pebble conglomerates (Figure 8.7), raindrop imprints, and salt casts. Caution must be used in interpreting mudcracks, because shrinkage polygons can form within a mass of sediment in colloidal layers that lose water during compaction and lithification. How could you distinguish such "dewatering" cracks that form within a sediment from desiccation cracks caused by drying in the sun?

Based on the knowledge of depositional environments that you have acquired from the previous chapters, you should now be able to predict which environments should be characterized by subaerial exposure, evaporation of water leaving salt crystals, and various types of water flow. For example, what kind of environment is characterized by water flowing alternately in opposite directions? You can find out more about specific environments from your instructor and from the background readings listed at the end of this chapter.

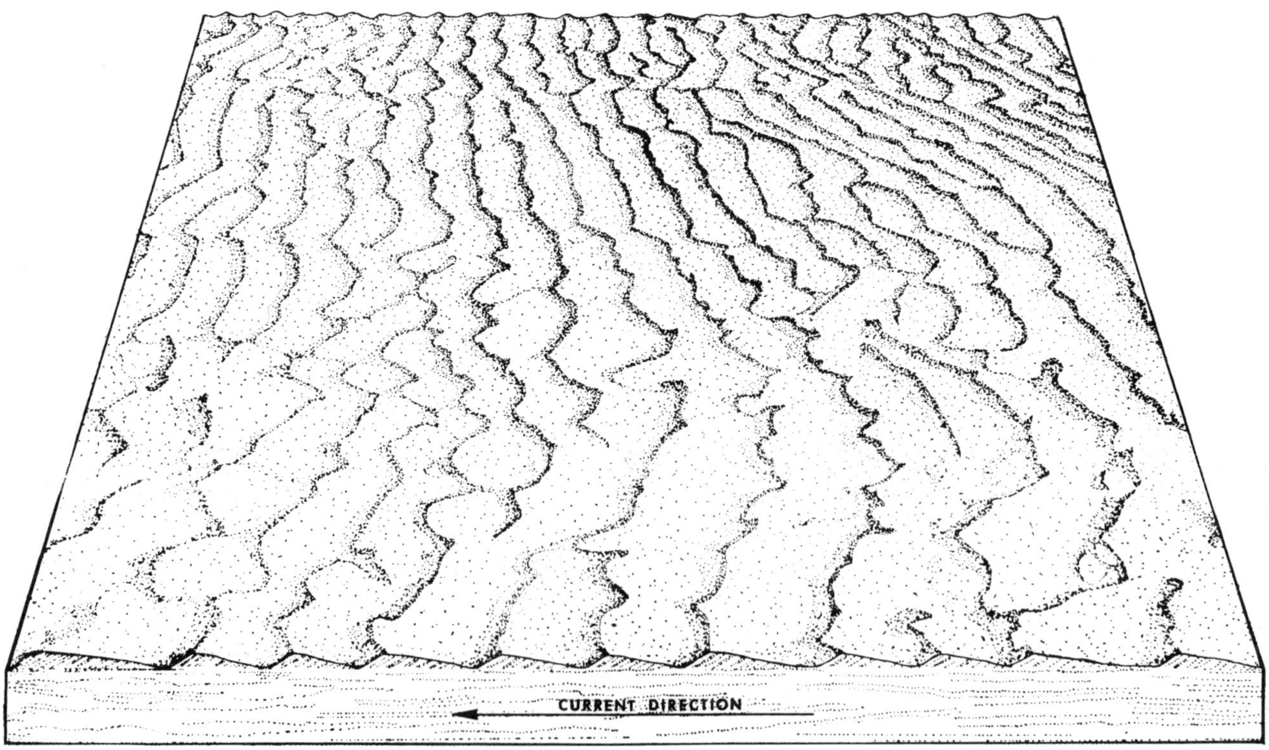

Figure 8.4. Low-velocity current ripples from the Chaco River, New Mexico. 15 CM

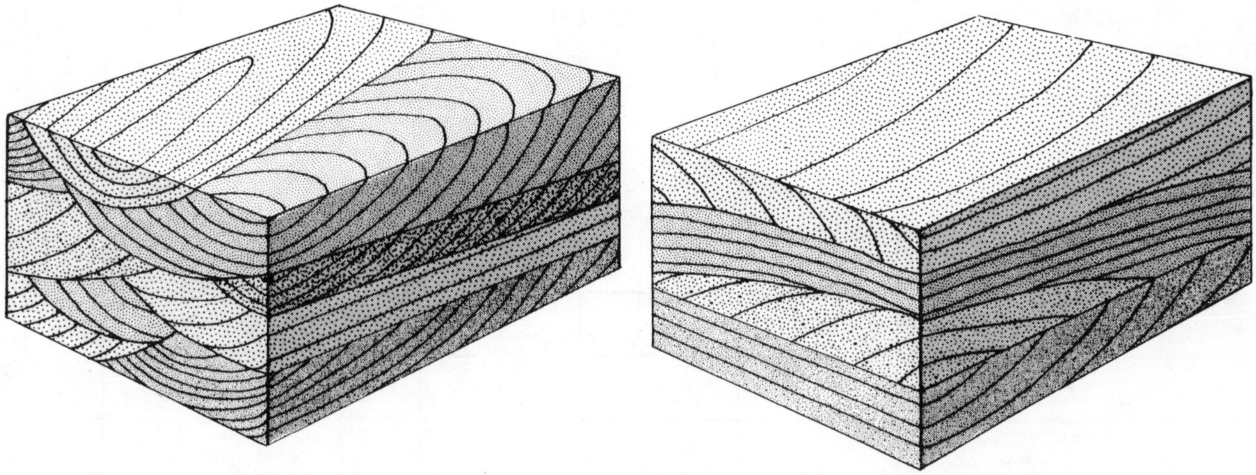

A. Festoon cross bedding B. Tabular cross bedding

Figure 8.5. Two types of superimposed cross-bed sets.

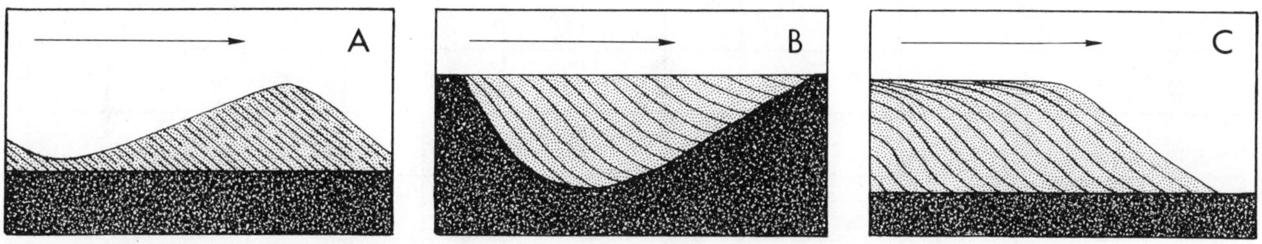

A. A small ripple ridge or large sand dune B. A scour-and-fill structure C. A point bar or lacustrine delta

Figure 8.6. Sedimentary structures showing internal cross bedding.

Exercise

1. Fill in the table by putting one X in a square if the sedimentary structure should be present and two X's in a square if the structure should be common.
2. As many of the structures listed in the table as are available will be set out in the laboratory, both as knowns and as unknowns.
   a. Identify each unknown and determine its most likely environment of deposition.
   b. Examine the other features of the rock, including grain size, roundness, sorting, organic content, and fossil types. Be sure to use the microscope. Does this information help you to reach more definite conclusions?
   c. Based on your examination of natural suites of sedimentary structures, make whatever corrections might be needed in the table. Why were some of your earlier assumptions possibly incorrect?

Questions for Discussion

1. Sediment grains (and sometimes fossils) are often eroded and redeposited as a new sediment that contains grains showing the effects of completely different processes and environments. Can the same thing happen to a sedimentary structure? Which is a more reliable guide, sedimentary structures or grain properties?

Environments / Structures	Alluvial fans	River floodplains	River point bars	Sand dunes (wind)	Lake bottoms	Swamps and marshes	Coastal lagoons	Marine beaches	Barrier islands	Estuaries and tidal flats	Open marine shelf
Flat, horizontal bedding											
High-angle cross beds											
Low-angle cross beds											
High-amplitude ripples											
Low-amplitude ripples											
Assymetrical ripples											
Oscillation ripples											
Sole marks											
Festoon cross bedding											
Tabular cross bedding											
Desiccation cracks											
Flat pebble conglomerate											
Edgewise conglomerate											
Round pebble conglomerate											
Raindrop imprints											
Salt casts											

2. Could field studies of sedimentary structures yield any kind of data that could be used to reconstruct part of the paleogeography of a large region?
3. What effects will burrowing and sediment-feeding organisms have on sedimentary structures? Might the presence of laminations, bedding, and other structures reflect the absence of sediment feeders? In this light, do sedimentary structures reflect high trophic stability or could they reflect an environment that is in some way unsuitable for benthic organisms?
4. Can limestones or dolomites have cross bedding, ripple marks, and mudcracks?

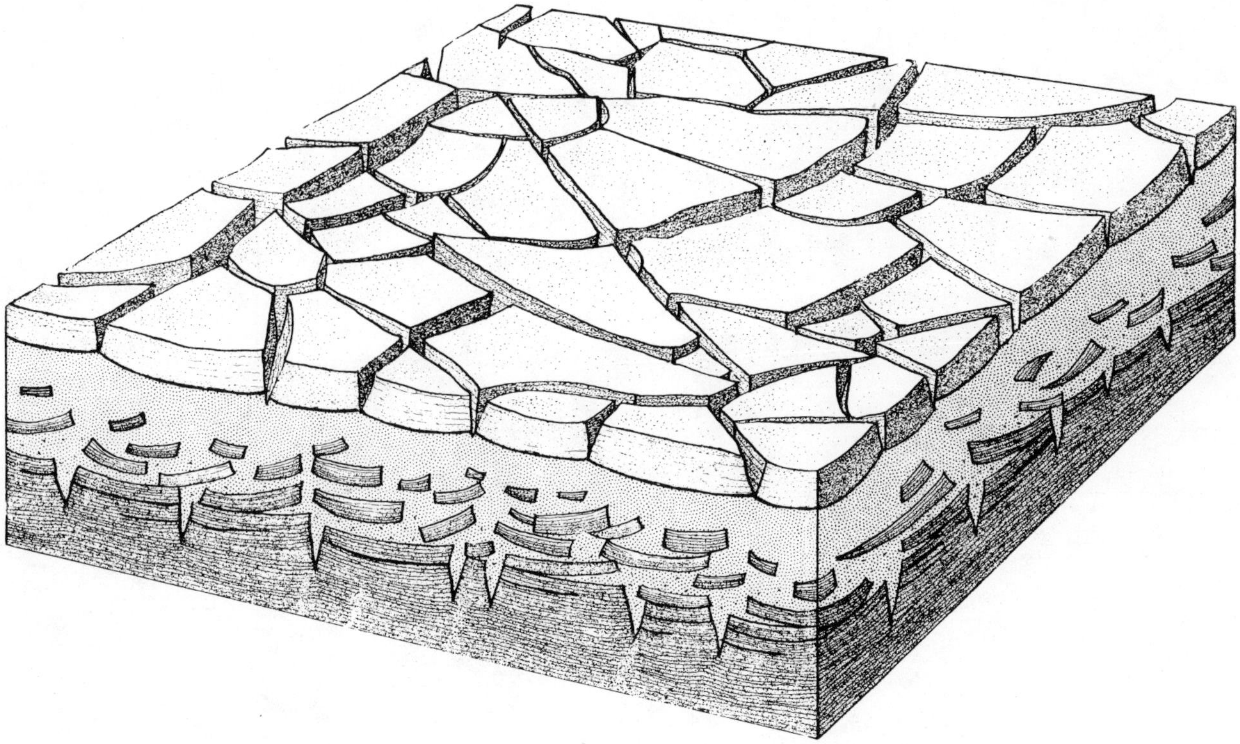

Figure 8.7. Mudcracks overlying a flat pebble conglomerate.

Background Reading

Blatt, H., Middleton, G., and Murray, R. 1972. Origin of Sedimentary Rocks, pp. 111-74. Englewood Cliffs, N.J.: Prentice-Hall.

Middleton, G. V., ed. 1965. Primary Sedimentary Structures and Their Hydrodynamic Interpretations. Society of Economic Paleontologists and Mineralogists, Special Publication 12.

Pettijohn, F. J., and Potter, P. E. 1964. Atlas and Glossary of Primary Sedimentary Structures. New York: Springer-Verlag.

Selley, R. C. 1970. Ancient Sedimentary Environments. Ithaca, N.Y.: Cornell University Press.

Chapter Nine

Paleontology
Trace Fossils

Objectives

1. To recognize the characteristics of sedimentary structures produced by the activity of animals working on the substrate.
2. To evaluate the usefulness of organic sedimentary structures in interpreting ancient environments.
3. To consider the importance of such structures in studying the evolution of organic communities.

Important Terms

Bioturbation structure—Any disruption of sediment by organic activity, including trace fossils without distinctive form, caused by intensive reworking of the sediment, sometimes producing a burrow-mottled texture.

Shaft—A vertical cylindrical burrow or the vertical part of a complex burrow system.

Spreiten—(singular, *spreite)* Bladelike or spiral series of many closely spaced parallel or concentric burrows produced by the migration of a single animal's burrow upwards, downwards, or sideways through the sediment.

Trace fossil—A sedimentary structure produced in either hard or soft sediment by the activities of an organism. Trace fossils include tracks, trackways, trails, shafts, tunnels, and borings.

Tracks—Isolated impressions left on the substrate by individual foot structures (podia).

Trackway—A continuous series of tracks made by the movement of a single animal.

Trail—A continuous groove or furrow left on the substrate surface by an animal that did not leave separate footprints but plowed through the sediment with all or part of its body.

Tunnel—A horizontal cylindrical burrow or the horizontal part of a complex burrow system. Tunnels can be distinguished from tracks because they are produced completely *within* the sediment.

Introduction

Trace fossils must be considered in a different light than fossil shells or skeletons, because they are not a former part of an animal but only a record of something the animal did. The same animal may produce several different types of trace fossils, depending on its behavior. Figure 9.1 shows crawling, foraging, feeding, and resting traces that all conceivably were made by the same type of trilobite. Conversely, many different kinds of animals (even from different phyla) may produce the same type of trace fossil if their burrowing or locomotive behavior is very similar. A U-shaped burrow (Figure 9.2A) may be formed as a dwelling burrow by a great variety of suspension feeders, including brachiopods, molluscs, annelids, and other wormlike phyla. Burrows made by the same animal may look slightly different in different environments, as Figure 9.2D shows. The same figure illustrates tunnels that may reflect sediment feeding (9.2B) and vertical dwelling burrows (9.2C).

Community Paleoecology

Common marine trace fossil assemblages include two groups of generally vertical dwelling burrows of suspension feeders, and three groups of more horizontal burrows and surface trails of sediment feeders (Figure 9.3). The suspension feeders represent shallow shelf communities of stable suspended food resources, whereas the sediment feeders represent outer shelf and continental slope communities whose stable resources are buried in the sediment. Nearshore areas of fluctuating food supplies may also be dominated by sediment feeders, such as the areas of heavy mud deposition near a delta. In general, trace fossils can be ascribed to either sediment feeders or suspension feeders, and trace fossil assemblages can be used to interpret the trophic stability of ancient communities, either with or without accompanying skeletal fossils.

Spreiten burrows may reflect either an animal's gradual shifting of its burrow upward because of sedimentation, or programs of simple or highly efficient sediment mining. In shallow-water sands, the upward migration of the burrow may simply result from high sedimentation rates. In areas where sedimentation is slow, the burrow may be migrating downward or sideways to accommodate the growing animal as it becomes larger. Simple shafts are most commonly made by filter feeders, to serve as dwelling burrows. The deep, vertical burrows support the filterers and also afford protection from the predators that are more common in communities having a stable supply of suspended food. The burrow of an infaunal filterer functions in the same way as the exoskeleton of an epifaunal filterer; that is, it provides both support and protection. A variety of shaft burrows are found in the shallow-water environments around a modern barrier island (Figure 9.4).

On the outer shelf and in deeper-water environments, much less food is suspended in the water, mostly owing to the lack of light penetration in such deep water and the absence of primary producers. Because of either less food in general or less stability in the food supply, fewer predators are found in these communities, a condition that could be reflected in the extensive development of surface sediment grazing traces. In communities having more predators or higher rates of sedimentation, the sediment feeders burrow more deeply, either with helical spreiten burrows or complexly branched tunnels and shafts (Figure 9.3).

Unfortunately, the abundance of recognizable burrows does not necessarily reflect the abundance of the former burrowers. In environments where extensive and deep sediment mining took place, the sediment may have been fairly well homogenized by organic reworking and lithified to a massive, somewhat structureless, rock. Burrows are best preserved where the sediment feeders are not prolific.

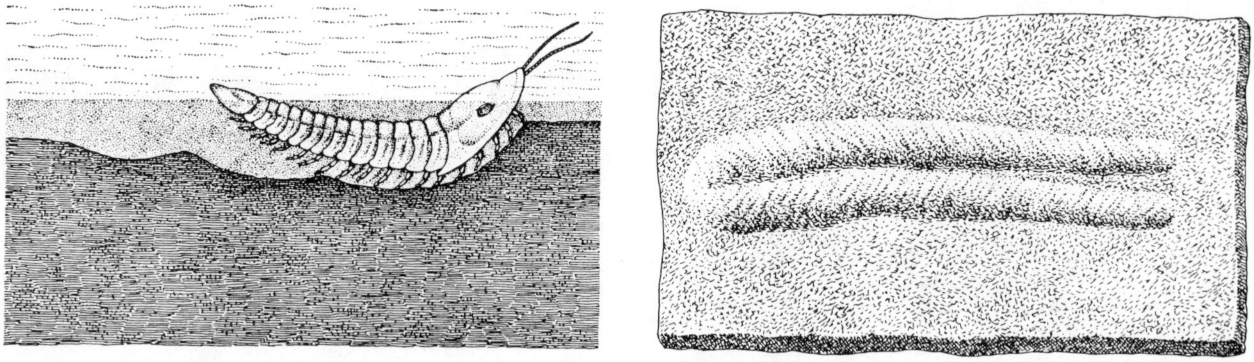

A. Crawling trail

B. Foraging trail

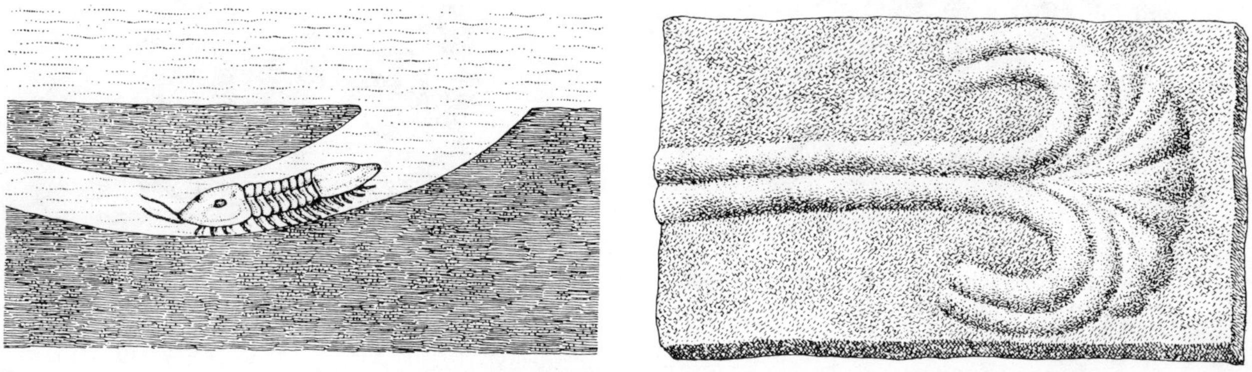

C. Feeding burrow

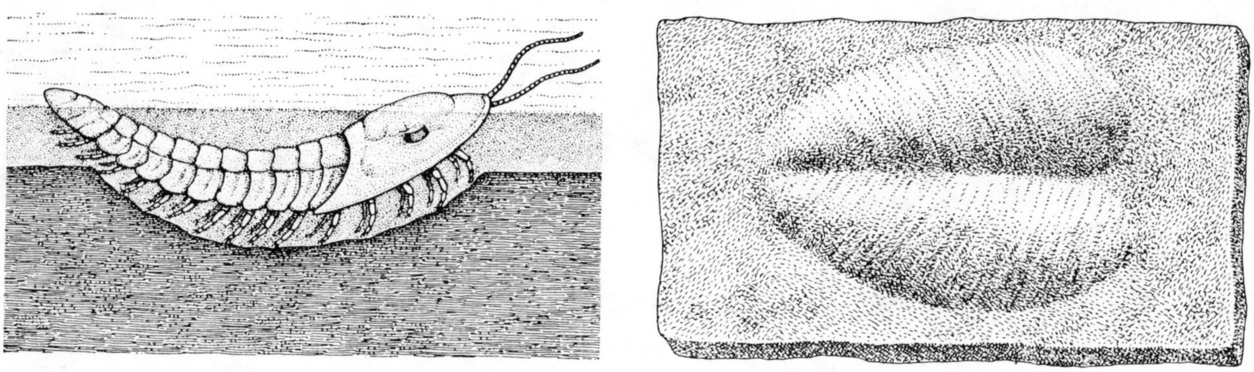

D. Resting trace

Figure 9.1. Different types of trails and burrows, possibly made by a single trilobite.

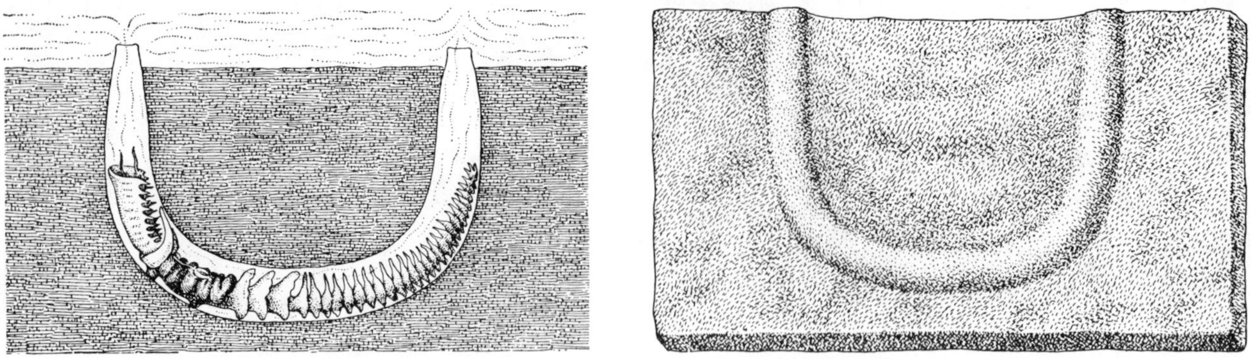

A. Dwelling burrow of a marine annelid. From "Fossil Behavior," Adolf Seilacher. Copyright © 1967 by Scientific American, Inc. All rights reserved.

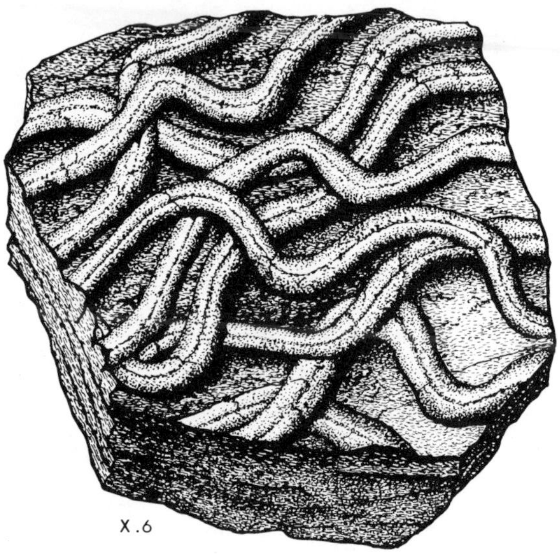

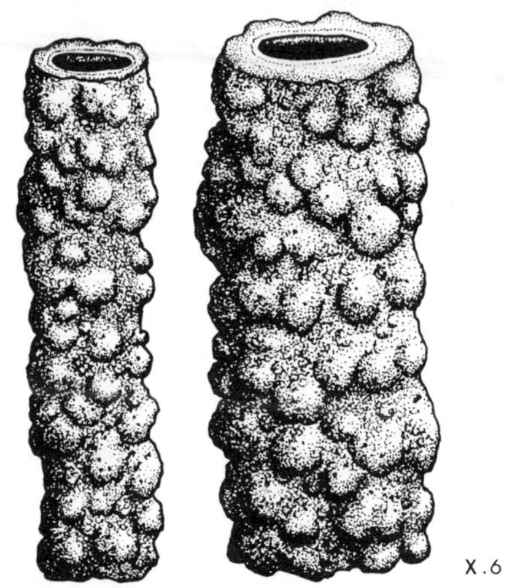

B. Slab of Ordovician limestone with the tunnels made by a wormlike animal.

C. Sections of knobby-walled shrimp burrows from the Miami Oolite, Florida Keys.

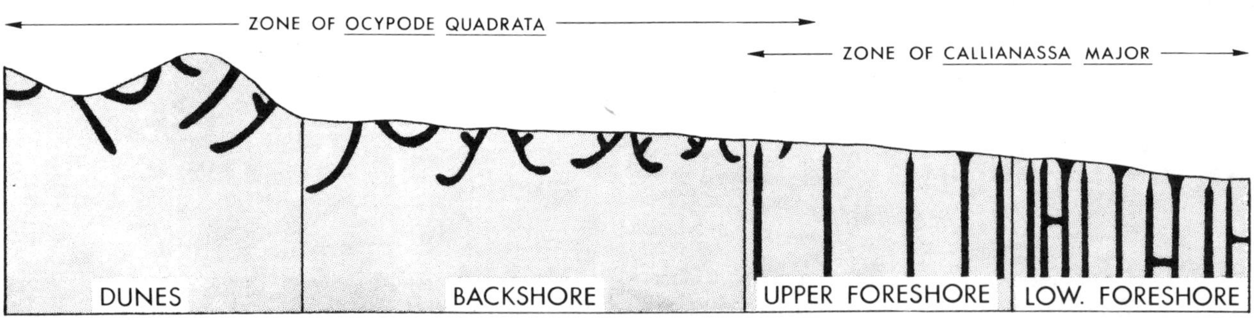

D. Shafts made by the ghost crab, *Ocypode,* and the ghost shrimp, *Callianassa,* in the beach-dune complex of Sapelo Island, Georgia.

Figure 9.2. Several types of burrows.

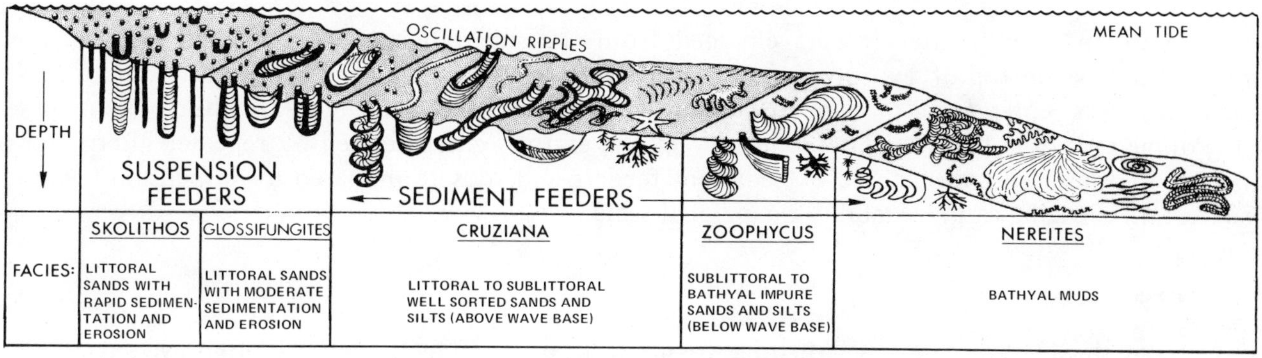

Figure 9.3. Trace fossil assemblages.

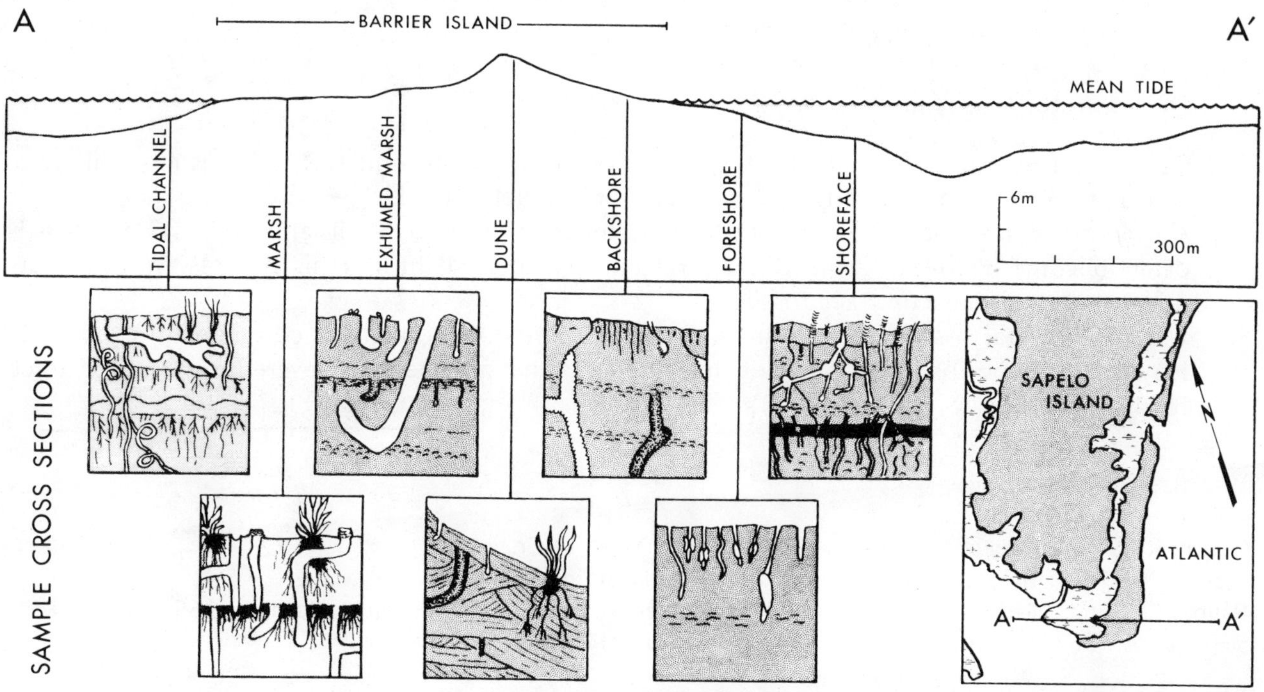

Figure 9.4. Variation in shaft burrows on the margin of a modern barrier island (Sapelo Island, Georgia).

Community Evolution

Because of their usefulness in interpreting the trophic stability of ancient communities, trace fossils are important in studying the evolution of community structure through geologic time. Burrowing is accomplished in unshelled animals when the pressure of fluids in the body cavity (coelom) makes part of all of the body rigid, and therefore capable of poking down into the sediment. Because fossil burrows are first found in late Precambrian rocks, about 700 to 750 million years old, and skeletal fossils are first found in Cambrian rocks, about 600 million years old, it is reasonable to suppose that the coelom developed in late Precambrian time before the widespread appearance of skeletons. The coelom enabled marine organisms to withstand fluctuations in trophic resources by making available the stored food in the sediment. Therefore, stable resources in the sediment allowed communities to diversify, and the coelomate body structure allowed the differentiation of all the higher animal phyla. When global resources became slightly more stable in the Cambrian period, filter feeders emerged from the substrate and developed exoskeletons as pseudoburrows on top of the substrate. The relatively sudden appearance in the Early Cambrian of skeletal fossils representing most of the animal phyla could have been due to a slight global increase in trophic stability. The Phanerozoic history of the earth is characterized by repeated alternations between communities dominated by sediment feeders and those dominated by suspension feeders, suggesting changing trophic stability on a global scale.

Exercise

Examples of different types of trace fossils will be set out as known and unknown specimens.

1. Study all the known specimens, and identify each unknown. Based on *all* available information, determine the depositional environment of each unknown as precisely as possible.
2. If any large slabs are available that contain a variety of trace fossils and skeletal fossils, determine all of the feeding types present in each and count their relative proportions. Rank the assemblages according to trophic stability.

Questions for Discussion

1. Can trace fossils be reworked and redeposited in other environments? Which is more likely to be exactly in its original habitat, a trace fossil or a skeletal fossil?
2. *Diagenesis* is the process whereby a sediment becomes lithified. Shells and other skeletal fossils often become dissolved, mineralized, or somehow altered during this process. What effects might diagenesis have on trace fossils?
3. Why do skeletal fossils provide an incomplete picture of animal (or community) evolution? How might trace fossils supplement that picture? What are the most severe limitations of trace fossils in studying evolution?

Background Reading

Cowen, R. 1976. History of Life, pp. 26-29, 43-45. New York: McGraw-Hill.

Frey, R., ed. 1975. The Study of Trace Fossils. New York: Springer-Verlag.

Raup, D. M., and Seilacher, A. 1969. "Fossil Foraging Behavior: Computer Simulation." Science 166:994-95.

Seilacher, A. 1967. "Fossil Behavior." Scientific American 217: 72-80.

Chapter Ten

Paleontology
Algal Stromatolites

Objectives

1. To study the origin and morphology of different types of algal stromatolites.
2. To evaluate their usefulness in interpreting ancient environments.

Important Terms

Algal mat—A dense, sheetlike mass of intertwined algal filaments, which may contain green algae and bacteria as well as blue-green algae.

Blue-green algae—Procaryotic photosynthesizing one-celled organisms, preserved as fossils in very ancient rocks and responsible for forming stromatolites.

Photosynthesis—The biochemical process whereby sunlight is used to produce food.

Stromatolite—A laminated sedimentary structure produced by blue-green algae, or possibly bacteria. Common overall forms are sheets, pillars, and spheres.

Introduction

Stromatolites are very thinly laminated three-dimensional masses of lithified lime mud. They are *neither* skeletal fossils *nor* trace fossils, but are sedimentary structures produced by a combination of organic and inorganic processes. They are mostly formed by the deposition of lime mud on top of, within, or underneath, an algal mat. A single algal mat may contain numerous different genera and species of living algae, up to 28 species having been identified in a single mat in southern Florida. The depositional environment also affects the shape of the mat, regardless of which species make it up.

Stromatolites are simply formed by the entrapment of sediment on the mucilaginous sheaths of the algal filaments, as illustrated in Figure 10.1. In some specimens, particularly those composed of silica rather than lime mud, the algal filaments are preserved as actual fossils within the stromatolite. The very thin daily-deposited layers of sediment may be flat, gently undulating, strongly arched, updomed, or columnar. They may also be spheroidal in mats coating loose nuclei

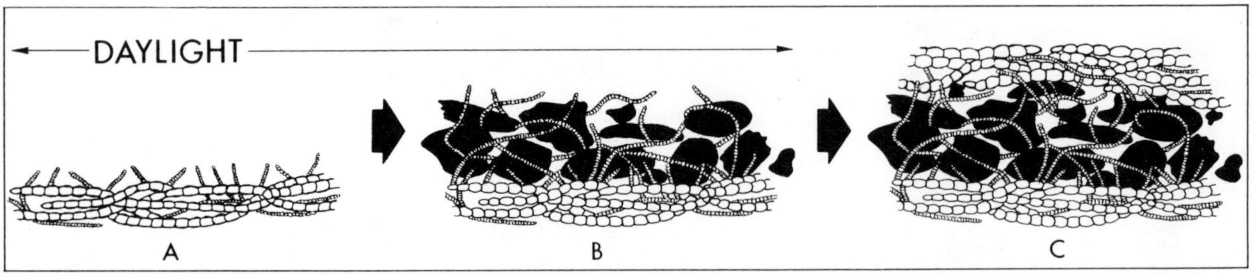

Figure 10.1. The entrapment of sediment particles by an algal mat; dark masses are sediment grains. The algae and sediment are highly magnified. (A) Uncovered mat at beginning of daylight period. (B) Sediment trapping during daylight. (C) Regrowth of mat and sediment binding during darkness.

that were free to roll around on top of the substrate. The layers of sediment simply reflect the shape of the algal mat on which the sediment was captured.

Origin of the Lime Mud

Algal stromatolites generally incorporate very little detrital mud or sand. It seems probable that the lime mud (or other chemical sediment) originates in the immediate vicinity of the algal mat and that biochemical processes may be the cause. At least two different mechanisms are possible.

Photosynthesis

The photoperiod (time of continuous daylight) controls the duration and intensity of photosynthesis, in which the following reaction takes place.

$$CO_2 + H_2O + \text{light} \rightarrow C_n(H_2O)_n \text{ (sugar)} + O_2$$

At night (in the presence of oxygen), this equation is reversed, and stored food is consumed.

$$C_n(H_2O)_n + O_2 \rightarrow CO_2 + H_2O + \text{energy}$$

Geologists have speculated that during the early Precambrian period, the burial of unoxidized organic matter, as well as food consumption through anaerobic fermentation, resulted in an excess of oxygen from photosynthesizers, which gradually built up our present oxygen-rich atmosphere. The accumulation of substantial oxygen allowed the development of the much larger eucaryotic cells, which first appeared about a billion years ago. Furthermore, the buildup of oxygen gradually produced a layer of ozone in the upper atmosphere, eventually shielding the earth's surface from most of the sun's ultraviolet radiation.

During daylight, according to the reactions above, carbon dioxide (CO_2) is removed from the water overlying and within the mat, but during the night CO_2 builds up in the water. Because dissolved CO_2 is a weak acid, calcium carbonate ($CaCO_3$) is more soluble in the water during the night than in the day. During daylight the water becomes more alkaline, and solid $CaCO_3$ can be precipitated as mud-sized particles, both within the algal mat and on its upper surface, where it is entrapped by the sticky filaments.

Anaerobic Bacteria

Reducing conditions in the lower layers of an algal mat resulting from the bacterial decay of dead algae can produce ammonia and precipitate $CaCO_3$.

$$NH_3 + H_2O \rightarrow NH_4OH$$

$$2\,NH_4OH + Ca(HCO_3)_2 \text{ (dissolved bicarbonate)} \rightarrow (NH_4)_2CO_3 + 2\,H_2O + CaCO_3 \text{ (solid)}$$

In some algal mats there is a gradual change between pure mat and pure "algal" micrite, which could be accounted for by the bacterial destruction of the mat and its replacement by $CaCO_3$. Micrite may be produced only in small patches where the conditions are more favorable for bacterial growth. Some stromatolites are arched or updomed; these swellings in an algal mat could have been caused either by surface irregularities below the mat or by pockets of gas released during bacterial decay.

Bacteria alone can form stromatolites, and some structures similar to some Precambrian stromatolites are forming today by bacterial precipitation from the waters of hot springs. Some large oolites may be simply spherical stromatolites and could involve bacteria. Bacteria probably play a role in the origin of other chemical sediments such as chert (amorphous silica) and pyrite.

Stromatolite Morphology

Stromatolites are classified by means of simple geometric formulas that describe their morphology. The basic types are illustrated in Figure 10.2.

LLH—Laterally linked hemispheroids with continuous laminae. The domes may be close together (LLH-C) or widely spaced (LLH-S) and separated by flatter areas.

SH—Stacked hemispheroids forming columns or clublike heads separated by nonlaminated sediment. The domes may have a constant diameter (SH-C) or a variable diameter (SH-V).

SS—Spheroidal structures surrounding a nucleus that rolled freely on the bottom. The laminations may be concentric (SS-C), randomly discontinuous (SS-R), or inverted (SS-I).

Transitional forms may be labeled with a hyphen, such as LLH-SH. Smaller forms may be superimposed upon larger ones. For example, if the laminae of large SH columns are arched like LLH forms, then the whole stromatolite is labeled LLH/SH. If there is considerable variation of form within a single stromatolite, then a more complex formula will be required.

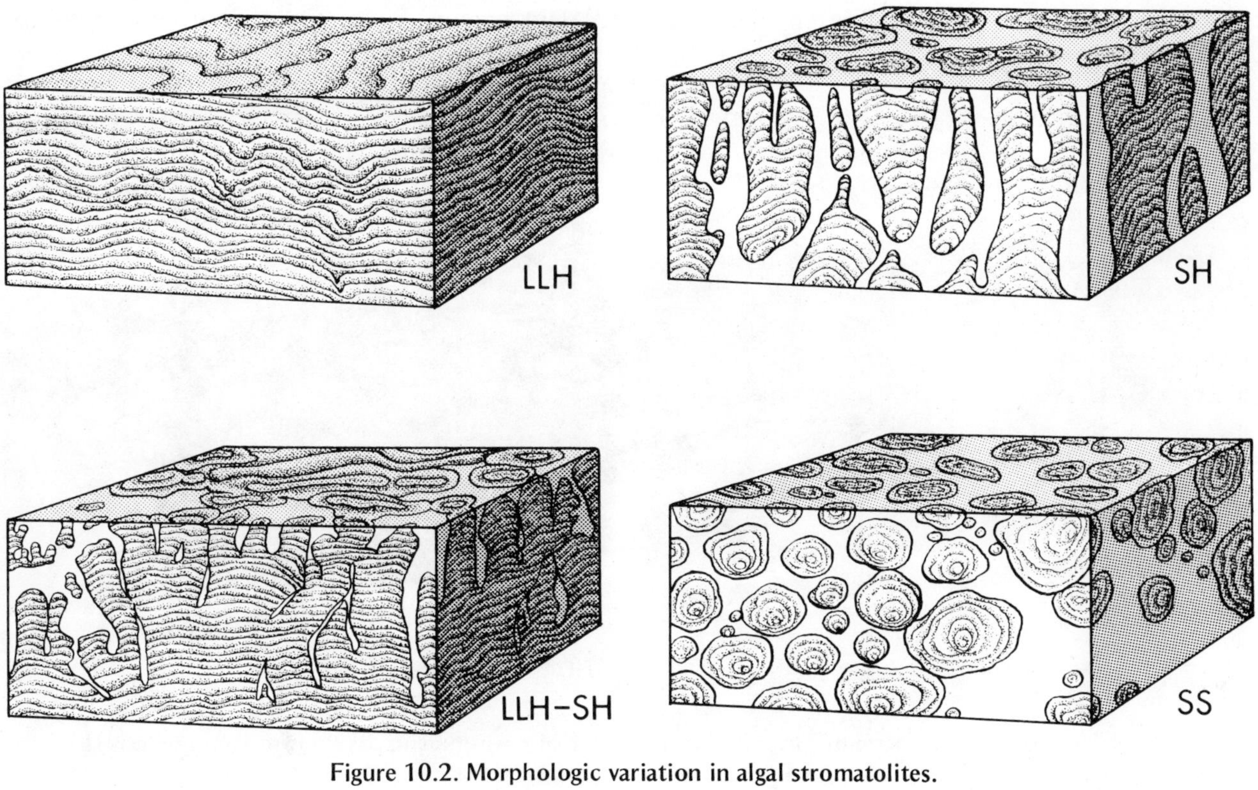

Figure 10.2. Morphologic variation in algal stromatolites.

Depositional Environments

Almost all stromatolites are found within or near the intertidal zone (Figure 10-3). The alternate wetting and drying in these environments allows the older layers to harden before new ones are added. Also, the algal mats grow most prolifically in the shallowest, most well lighted environments. In the supratidal zone and upper intertidal zone, the mats grow in continuous sheets and produce LLH stromatolites. Because dolomite (see Chapter 11) is so commonly formed in supratidal environments, many ancient LLH stromatolites are dolomitized. To cause the separate growth of the clublike SH stromatolites, tidal scour is necessary to gouge out the sediment in between the columns (Figure 10.4). Therefore, SH forms grow in areas of vigorous surf action, as in exposed headlands or open beaches. Early hardening of the rims of the SH types prevents the collapse of the tall columns of sediment. Type SS stromatolites are found only in permanently submerged shallow water or very low in the intertidal zone, where surf action keeps turning them over so that the lamination can grow as spheroids. Most modern stromatolites are forming only in warm-water regions where aragonite is precipitated from supersaturated seawater, especially in hypersaline lagoons or bays. They can also form along the margins of hypersaline lakes, as along the shorelines of the Great Salt Lake in Utah and salt lakes in western Australia. Other sedimentary features, desiccation cracks for example, may be helpful in specifying the exact depositional environment of stromatolites.

The Cropping Hypothesis

Algal stromatolites are the best-preserved representatives of the abundant producer-dominated communities of the Precambrian. In the Cambrian period, stromatolites became very reduced in abundance, as they still are today, precisely at the same time that the diverse animal communities of the Phanerozoic eon appeared. In fact, stromatolites today grow well only in hypersaline environments, habitats that are hostile to most metazoans. The questions to be asked are: (1) Did they always form in hypersaline environments? (2) Did grazing by metazoan herbivores remove

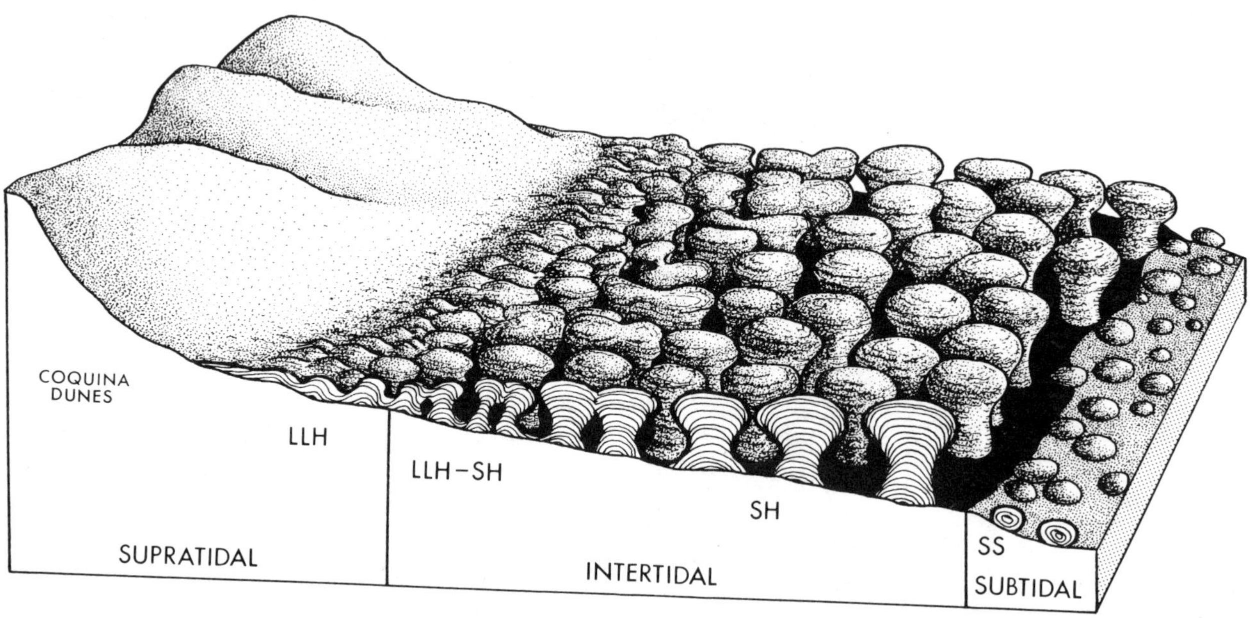

Figure 10.3. Differences in stromatolites in different depositional environments, as known from modern stromatolites in the tidal environments of a hypersaline lagoon in western Australia.

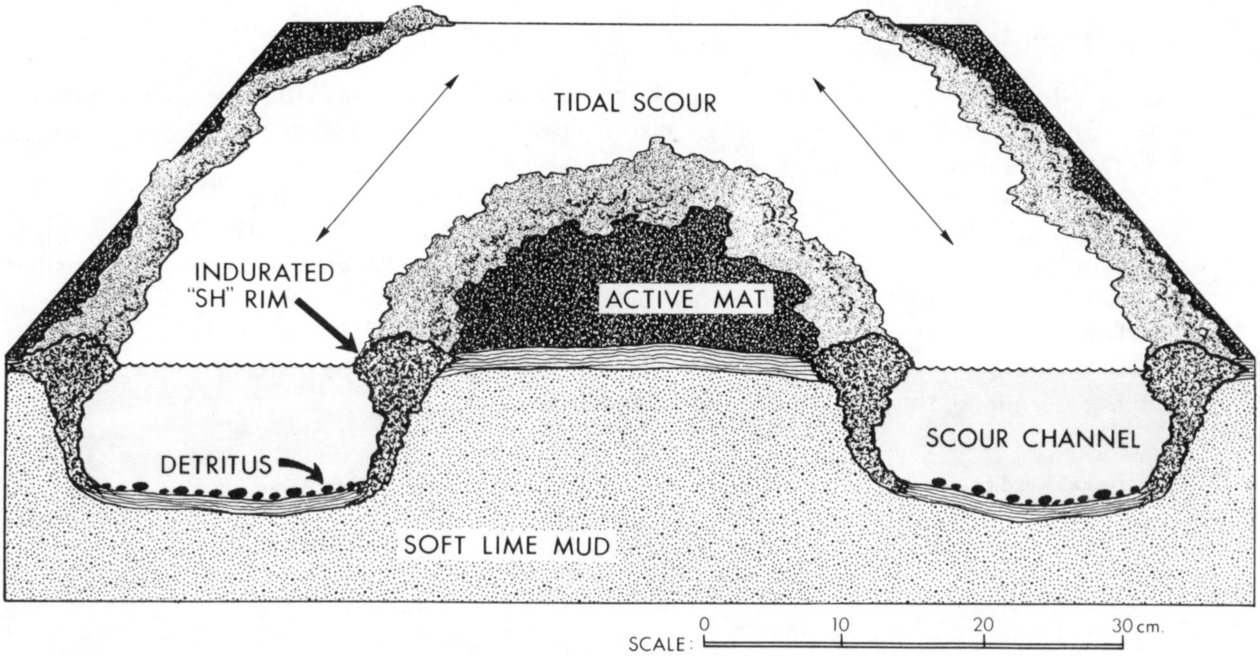

Figure 10.4. Origin of clublike stromatolites (SH) by tidal scour and rapid induration (hardening) of the rim.

them from normal marine environments? and (3) Was their reduction in such environments partly responsible for the relatively sudden appearance of diverse animal communities at the beginning of the Phanerozoic eon?

Living communities made up primarily of photosynthetic organisms tend to be dominated by only one or two species, which proliferate to such an extent that they monopolize all available space. Experiments have shown that a herbivore, or cropper, introduced into such a community, will not lower its diversity but will greatly increase it. The reduction of the dominant species allows a diversity of other species to grow, and they, in turn, permit a wider variety of herbivores and scavengers, which in their turn support a diversity of carnivores. The cropping principle is that reducing the domination of one or two prolific food species leads to the diversification of the entire community.

Because of the absence of Precambrian metazoans, the algal communities went uncropped for over 2 billion years. The evolution of an otherwise unknown cropper near the beginning of the Cambrian period could have caused the great reduction of algal stromatolites accompanied by the relatively sudden diversification of marine communities. This simple hypothesis could also explain why stromatolites are found today only in restricted environments.

Exercise

Examples of known and unknown algal stromatolites will be set out in the laboratory.

1. Identify all of the unknown stromatolites by comparing them to known specimens and to Figures 10.2 and 10.3.
2. Using *all* available information, determine the depositional environment of each specimen or specimen suite.
3. If any other fossils, either skeletal fragments or traces, are associated with the stromatolites, identify them and estimate the proportions of feeding types, including photosynthesizers, in each community represented. Bear in mind that some fossil fragments could have been brought in from other environments. How do these communities compare with others previously studied with respect to diversity and trophic stability?

Questions for Discussion

1. Explain why stromatolites should not be referred to as "fossil algae." In some older literature, stromatolitic deposits were commonly called "reefs." Are stromatolites reef deposits, and are the organic communities similar?
2. Why would stromatolites be particularly useful in paleogeographic mapping?
3. The height of columnar stromatolites is related to the amplitude of the tidal zone. How might the geomorphology of a coastline affect the height of tides? Would the analysis of ancient stromatolites be of any use in studying the history of the moon?
4. The laminations in stromatolites are deposited on a daily basis. Based on the discussion on growth banding in Raup and Stanley (1971, pp. 353-57), discuss how stromatolites might be used in determining the absolute age of ancient rocks.

Background Reading

Raup, D. M., and Stanley, S. M. 1971. Principles of Paleontology, pp. 215-17. San Francisco: W. H. Freeman.
Walter, M. R., ed. 1976. Stromatolites. Amsterdam and New York: Elsevier Scientific Publishing Company.

Chapter Eleven

Sedimentary Petrology
Origin of Dolomite

Objectives

1. To determine, as precisely as possible, the succession of depositional environments in a vertical sequence of dolomitized sedimentary rocks.
2. To determine the lateral distribution of the ancient environments from the vertical sequence, using a principle of stratigraphy known as Walther's law.
3. To determine the environment in which the dolomite originated and the model for the origin of dolomite that best fits the rock sequence.

Important Terms

Dorag dolomitization—Dolomitization by a mixture of seawater and fresh water.

Supratidal dolomitization—Dolomitization in supratidal sediments in which surface evaporation causes landward flowage of seawater through the sediment.

Walther's law—The idea that the vertical succession of rock types observed at a single locality reflects the lateral succession of environments that have migrated over that depositional site. This is a general interpretative principle and not a law in the strict sense.

The Supratidal Model

Two areas where dolomite is forming today are the Abu Dhabi sabkha in the Arabian Gulf (Figure 11.1) and Andros Island in the Bahamas (Figures 7.6 and 11.2). In both areas, evaporation in the supratidal zone causes seawater to flow upward to the surface, where evaporation causes the precipitation of gypsum, $CaSO_4$ (2 H_2O). As a result of the precipitation of the gypsum, the ratio of magnesium to calcium in solution rises, allowing the replacement of calcium carbonate by dolomite according to the reaction

$$Mg^{++} + 2\,CaCO_3 \rightarrow CaMg(CO_3)_2 + Ca^{++}$$

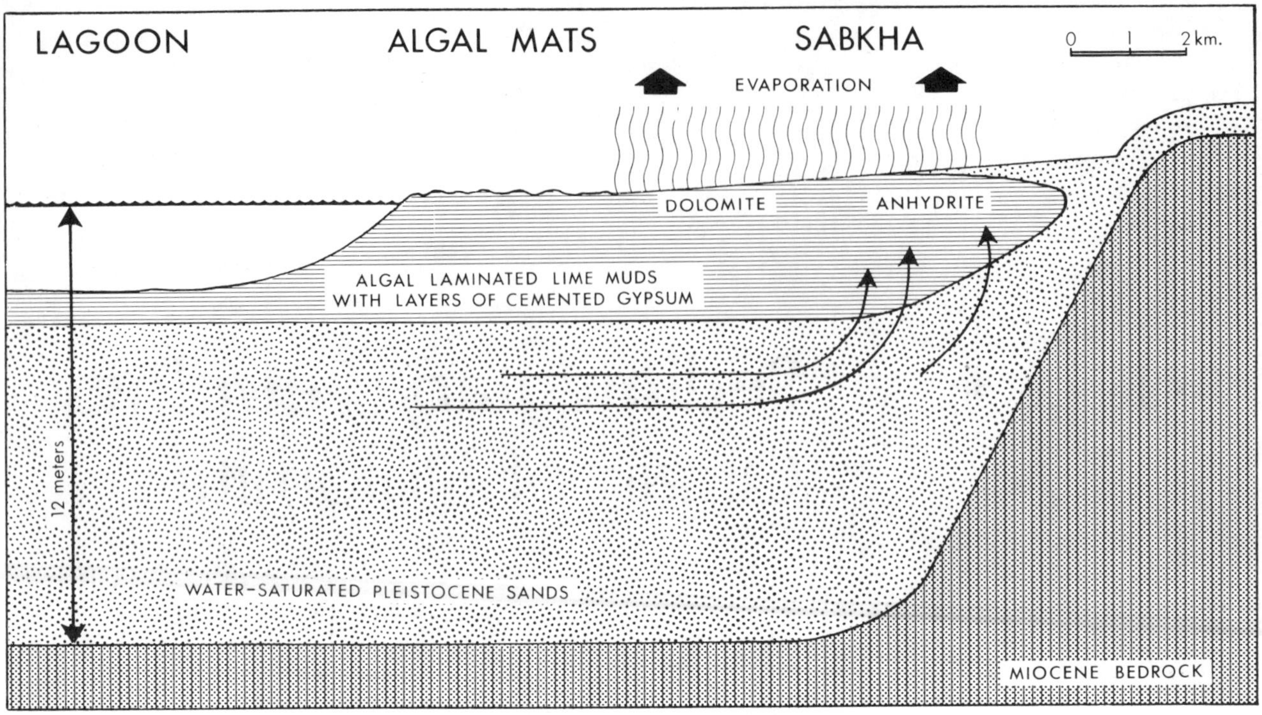

Figure 11.1. The Abu Dhabi sabkha in the Arabian Gulf

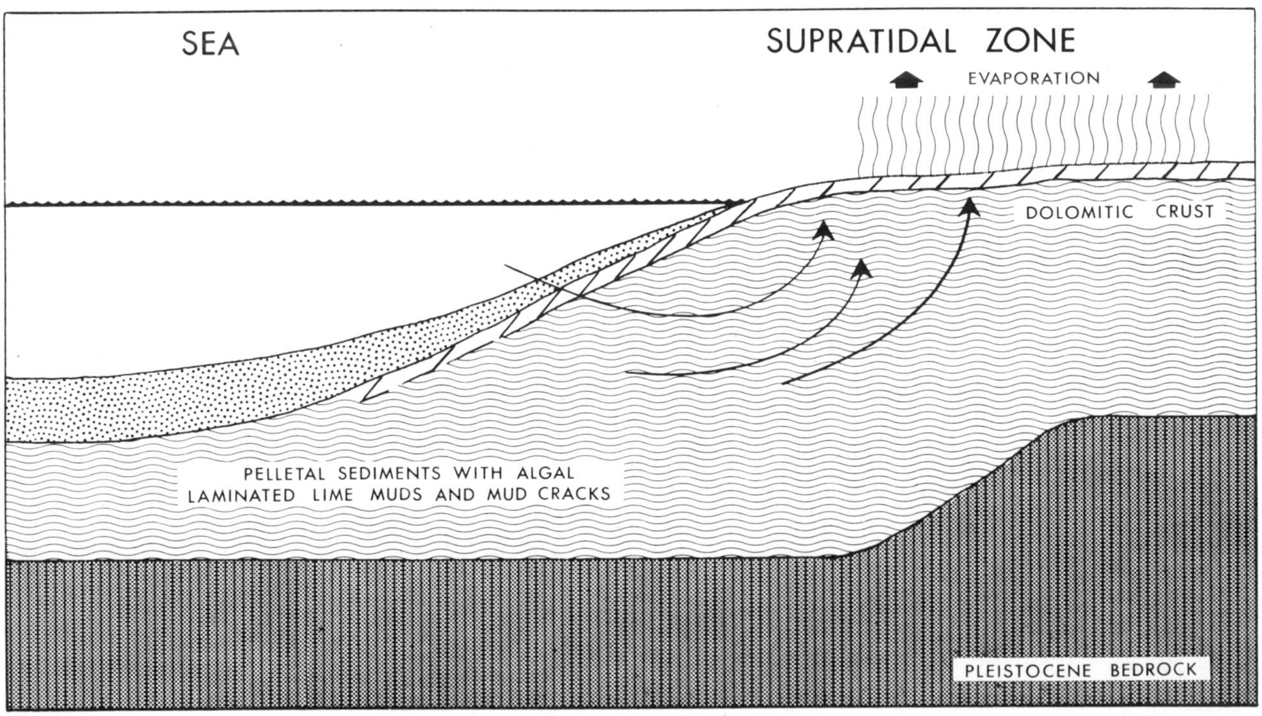

Figure 11.2. The dolomitic crust on Andros Island, Great Bahamas Bank; seaward the crust is older, having been buried as a result of the rise of sea level. The thickness of the crust is exaggerated; it is actually not thicker than 3 centimeters.

The original sediment, prior to dolomitization, may have been pelletal lime mud, as in the Bahamas, or algal stromatolites, as in the Arabian Gulf. Because dolomitization involves dissolution of calcium carbonate followed by precipitation of dolomite, original textures, including fossils, are often destroyed during the process. However, such rocks often have sedimentary structures characteristic of the supratidal zone, including LLH stromatolites, mudcracks, and chert nodules (silicified gypsum or anhydrite). A reconstruction of an ancient supratidal dolomite environment is illustrated in Figure 11.3.

The Dorag Model

Alternatively, dorag or mixed saline-fresh waters can flow through rocks and cause dissolution of calcium carbonate and precipitation of dolomite. This process is presently taking place in Jamaica near the shoreline, where fresh water in the ground mixes with seawater (Figure 11.4). The major difference between this model and the supratidal model is that dorag dolomitization is not restricted to a specific environment of deposition; it may affect intertidal and subtidal sediments. One characteristic of dorag dolomitization is the presence of cavities lined with limpid dolomite crystals. These crystals are often about 0.1 millimeters in diameter and are easily recognized with a hand lens or low-power microscope.

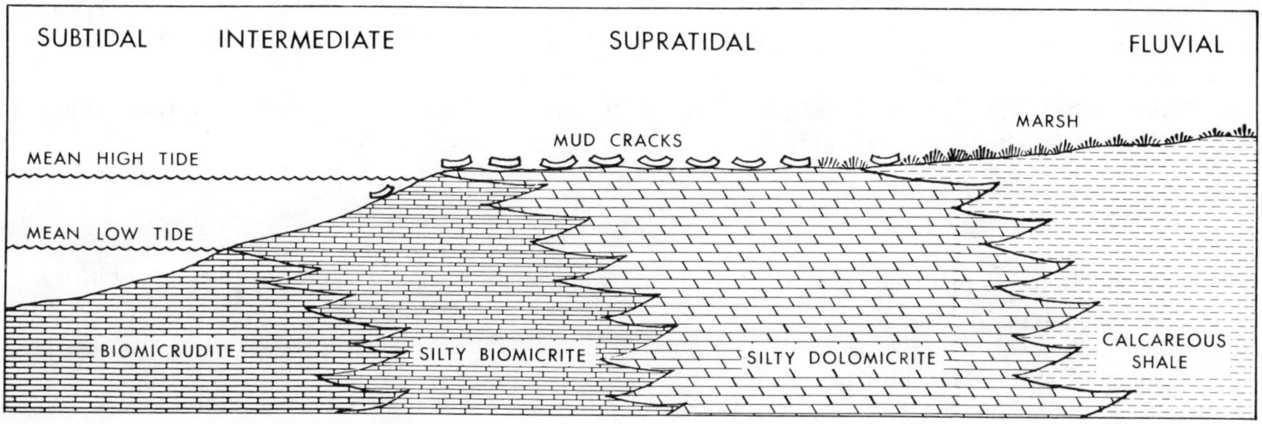

Figure 11.3. Inferred lateral distribution of environments associated with supratidal dolomite in the Upper Ordovician rocks of the southern Appalachians.

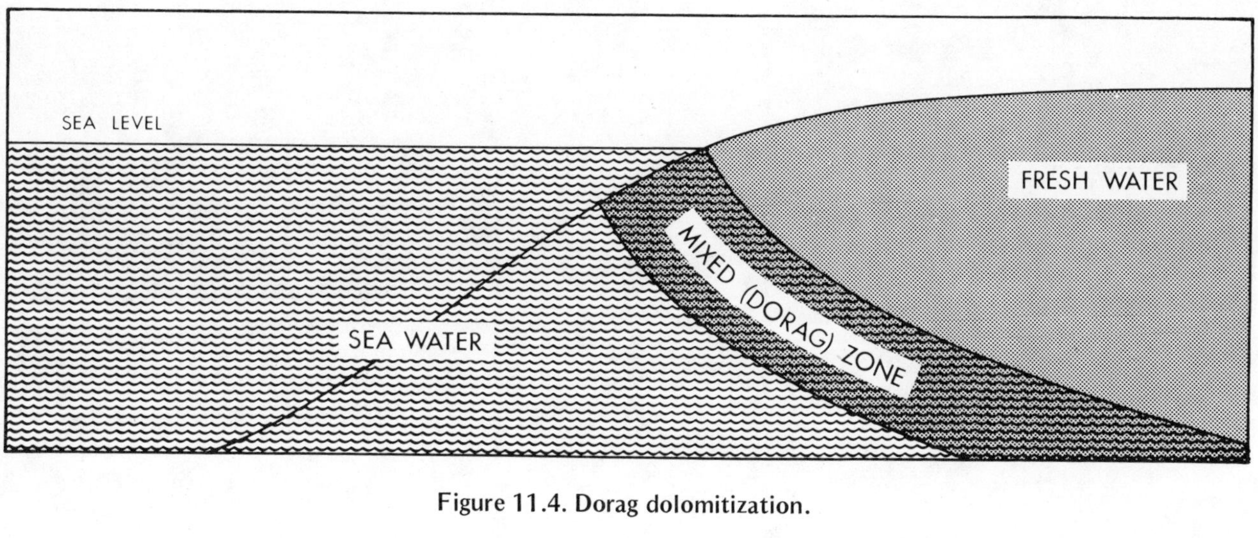

Figure 11.4. Dorag dolomitization.

Exercise

A collection of samples representing the vertical succession of rock types at a single locality will be set out in the laboratory. The rock samples containing dolomite will be identified and so labeled by your instructor. Simple acid tests are not reliable; thin section work, staining, and other methods are more conclusive.

1. Describe the sediment type, fossils, and sedimentary structures in each sample. Determine the depositional environment of each sample as precisely as possible, using all available information.
2. Using Walther's law, determine in which environments dolomitization took place and to what degree. Using Figure 11.3 as a model, make a diagram of the lateral distribution of environments.
3. Which of the two models (supratidal or dorag) appears to provide a better explanation for the origin of the dolomites you have studied?

Questions for Discussion

1. What sedimentary structures and fossils should be associated with supratidal sediments? With dorag dolomite? Which dolomites should be more closely associated with coral reefs? With algal stromatolites?
2. What climatic conditions are necessary for the development of dolomite? Is dolomite a paleoclimatic indicator?
3. Using the supratidal model, under what conditions would you expect to find dolomite stratigraphically above a coral reef facies? Below a coral reef facies?

Background Reading

Blatt, H., Middleton, G., and Murray, R. 1972. Origin of Sedimentary Rocks, pp. 477-500. Englewood Cliffs, N.J.: Prentice-Hall.

Badiozamani, K. 1973. "The Dorag Dolomitization Model-Application to the Middle Ordovician of Wisconsin." Journal of Sedimentary Petrology 43:965-84.

Butler, G. P. 1969. "Modern Evaporite Deposition and Geochemistry of Coexisting Brines, the Sabkha, Trucial Coast, Arabian Gulf." Journal of Sedimentary Petrology 39:70-89.

Chapter Thirteen

Stratigraphy Patterns of Shoreline Sedimentation

Objectives

1. To examine in greater detail the depositional environments in shoreline and coastal areas.
2. To determine the stratigraphic patterns that will result from the lateral migrations of these environments.
3. To determine, through simulation of a shoreline depositional system, the relationships of the resulting stratigraphic units to geologic time.

Important Terms

Isochron—A physical surface or layer that represents points of nearly simultaneous deposition within a body of sedimentary rocks. Examples are volcanic ash deposits (bentonites), varves in lake deposits, storm-deposited beds, lava flows, or the product of any depositional process that could *simultaneously* affect a fairly large geographic area.

Regression—The movement of onshore environments in the offshore direction, usually caused by falling sea levels. In regressive sequences, the most offshore environments are stratigraphically lowest, and the most shoreward environments are highest.

Transgression—The movement of offshore environments in the onshore direction, usually caused by rising sea levels. In transgressive sequences, the most offshore environments are stratigraphically highest, and the most shoreward environments are lowest.

Time surface—An imaginary surface that passes through a body of rock, representing the loci of all sediment that was deposited at nearly the same time.

Introduction

Review the depositional environments shown in Figure 2.3. As an exercise, make a list for each environment of the range of possible sediment characteristics (grain size, sorting, organic content), skeletal fossils likely to be preserved, trace fossils, and inorganic and organic structures. In other words, synthesize all the information you have learned so far in this manual for these particular environments. Review your lists for the environments shown in Figures 13.1 and 13.2.

Consider now the effects of changes in sea level on deposition in these environments. If sea level gradually fell, all of these environments would migrate in a seaward direction and would stay in the same positions with respect to each other. Eventually the marsh, lagoon, and barrier island sediments would be deposited in areas that had once been a marine shelf (Figure 13.2). The lateral movement of these environments could generate continuous flat-lying bodies of sediment of fairly uniform lithology. In a regressive sequence the most seaward environments would form the lowest beds, and sediments from the shoreward environments would be layered above them in the same stratigraphic order as the environments were arranged geographically. This common phenomenon is known as Walther's law (see Chapter 11). It states that the *geographic* order of environments will be reproduced in the *stratigraphic* column if the environmental complex migrates in a simple fashion in one direction. We should not expect the vertical sequence of environments in a series of bedded rocks to differ from the horizontal sequence of environments that we see in a modern depositional system.

Exercise

Part I (Optional)

Rock samples or suites of rock samples collected from shoreline environments (different from those used in the preceding exercises) will be set out in the laboratory. Identify the depositional environment of each, using *all* of the information at your disposal. Use the lists that you made earlier and compare each sample suite to what would be expected at each deposition site.

Part II—Simulation of Shoreline Deposition

Place a full page of blank tracing paper over Figure 13.1 so that the left side of the paper is about 8 centimeters to the right of the right edge of the barrier island, and align the paper parallel to the barrier island. Trace the patterns of the upland, marsh, lagoon, barrier island, and marine shelf environments. Use a different color or pattern for each environment; divide the marine shelf into an inner detrital sediment area and an outer area of carbonate sediments.

In this exercise you will try to simulate environmental migrations that might take place over a short interval of geologic time, say several million years. Using sheets of tracing paper, you will move the map pattern of environments in both the offshore and onshore directions in an effort to simulate both regressive and transgressive sequences. Assume that the pattern traced on your base map, as described in the preceding paragraph, represents time 1. On successive sheets of tracing paper, show the following changes, using a scale to determine distances (on the map 1.05 centimeters equals 1.00 kilometers).

Time 2—Shoreline migrates 5 kilometers seaward.
Time 3—Shoreline migrates 5 more kilometers seaward.
Time 4—Shoreline migrates 3 kilometers landward.
Time 5—Shoreline migrates 5 more kilometers landward.

Draw a complete sediment map for each position of these environments in time. You should vary minor details of the depositional system from time to time, including the positions of rivers and streams, the width of tidal marshes, and barrier islands and the locations of tidal deltas and inlets. Assume that the amount of sediment deposited in each location reaches a constant level before the environment migrates either seaward or landward. Also assume that topographically high areas, such as the barrier island, may experience partial erosion before other, lower-lying environments can migrate over them. In general, your model should produce continuous, flat-lying beds of nearly uniform lithology. The thickness of each bed will depend upon the geographic width of the specific environment and the rate at which it is being displaced.

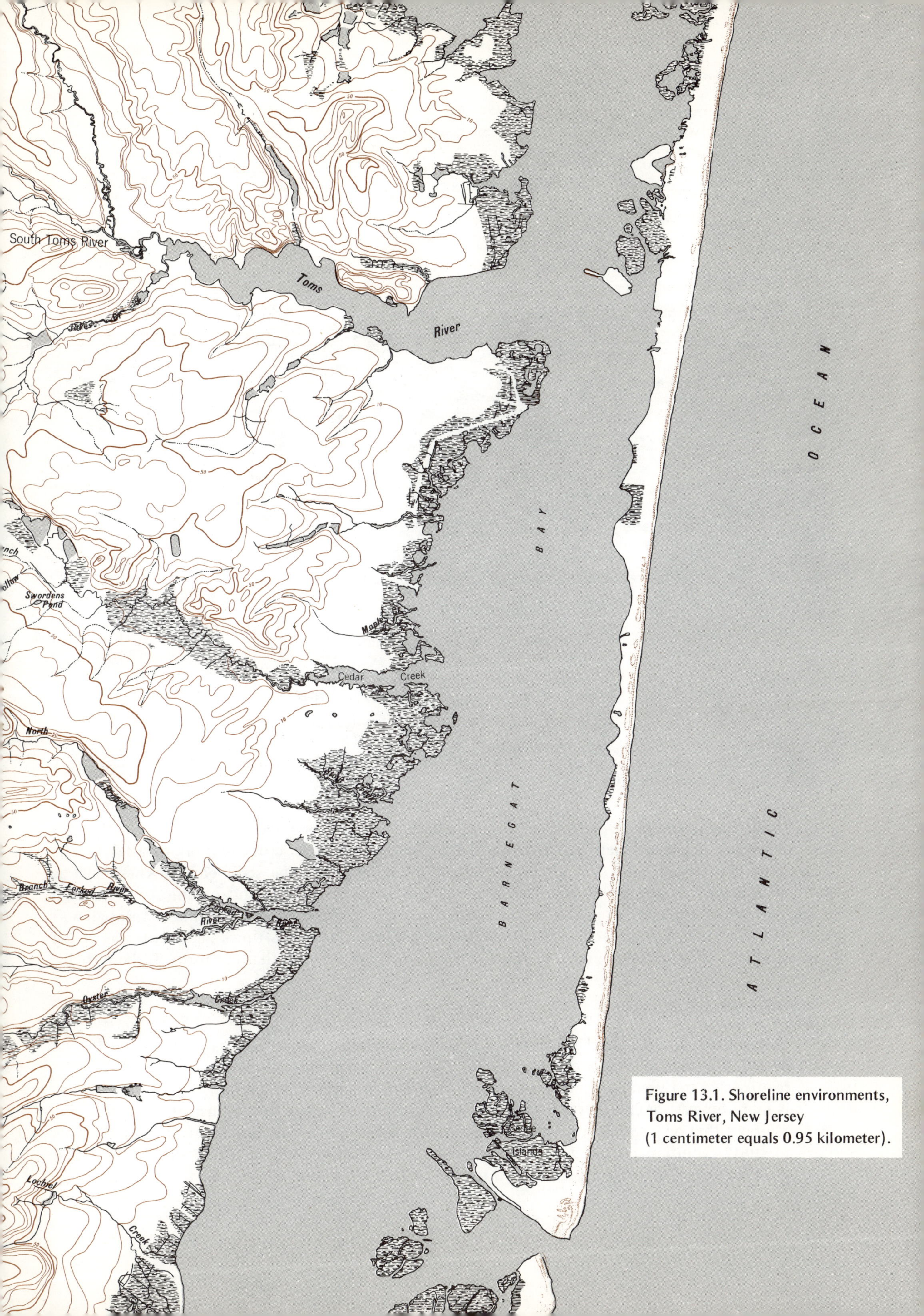

Figure 13.1. Shoreline environments, Toms River, New Jersey (1 centimeter equals 0.95 kilometer).

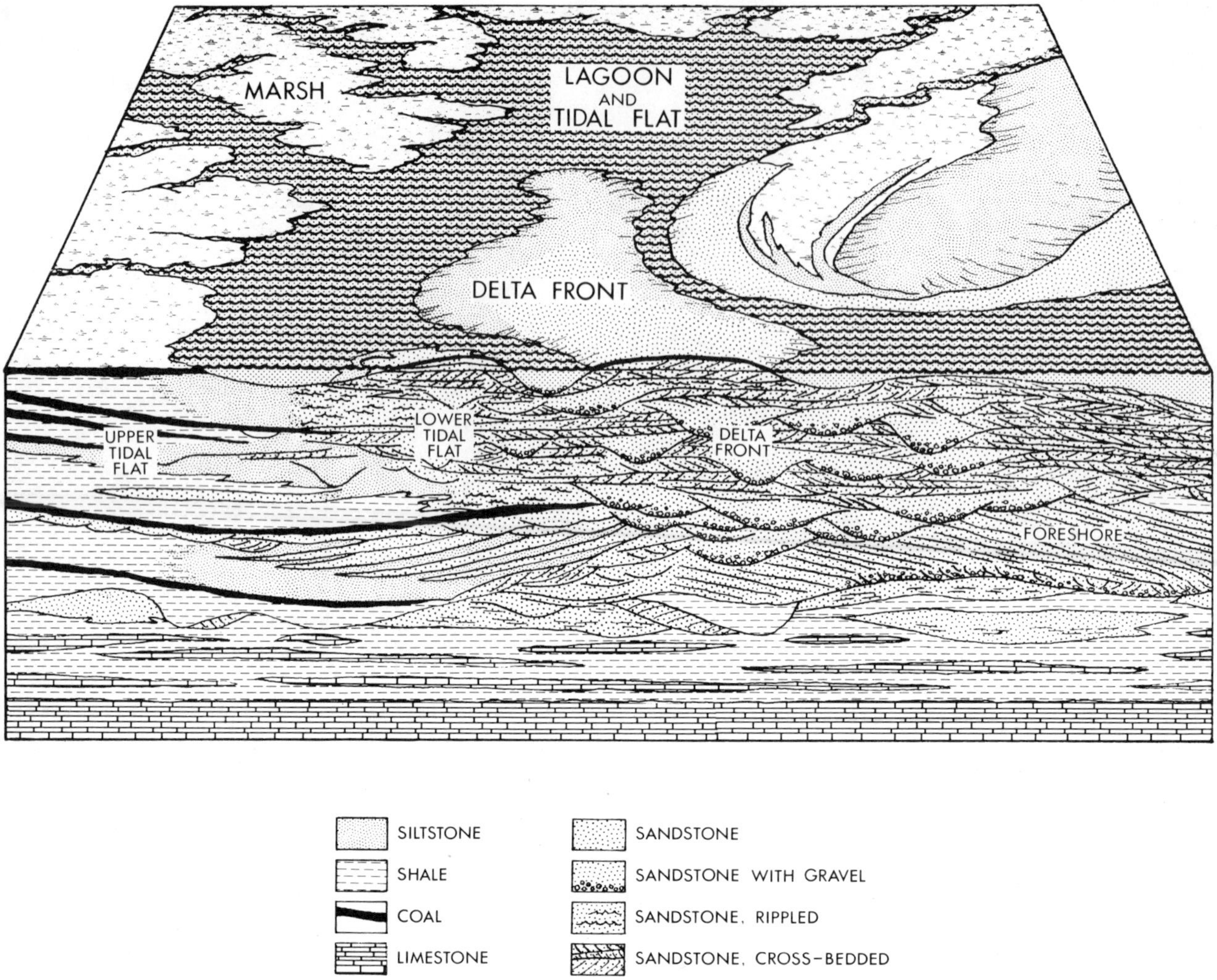

Figure 13.2. A regressive shoreline sequence with a small tidal delta. Notice that the onshore sediments overlie shales and carbonates of the open marine shelf.

Using the sediment maps for times 1 through 5, construct one vertical cross section through the sediments deposited perpendicular to the shoreline, and one cross section parallel to the shoreline. The vertical axis of your cross section should measure the sediment thickness, just as though you were taking drill cores at a series of locations. Show each unit of uniform lithology, using the same colors or patterns as on the maps. Draw in the time surfaces for times 1 through 5 as dark lines cutting across each section. Because you know at what times sediment was being deposited at what locations, you can draw in the exact time surfaces that cut through these rocks.

Questions for Discussion

1. You should notice a striking difference in the way time lines cut across or follow lithologic boundaries in your two cross sections. Can you formulate a general principle for the relationship between time and stratigraphy in shoreline sequences?
2. Is more time represented in the lateral or vertical direction in any homogeneous rock unit?
3. Suppose that an organic community inhabits areas of shelf carbonates. A paleontologist wants to study evolutionary changes in this community through some interval of geologic time. How should a sampling design be structured to maximize the duration of geologic time observed?

4. In any transgressive-regressive sequence of shoreline deposits, which time surface that can be easily identified represents an isochron?

Background Reading

Blatt, H., Middleton, G., and Murray, R. 1972. Origin of Sedimentary Rocks, pp. 206-9. Englewood Cliffs, N.J.: Prentice-Hall.

Dickinson, K. A., Berryhill, H. L., and Holmes, C. W. 1972. "Criteria for Recognizing Ancient Barrier Coastlines." In J. K. Rigby and W. K. Hamblin, eds., Recognition of Ancient Sedimentary Environments, pp. 192-214. Society of Economic Paleontologists and Mineralogists, Special Publication 16.

Eicher, D.L. 1976. Geologic Time, pp. 41-44. Englewood Cliffs, N.J.: Prentice-Hall.

Chapter Fourteen

Stratigraphy
Patterns of Fluvial and Deltaic Sedimentation

Objectives

1. To examine in greater detail the depositional environments in marine deltas and on river flood plains.
2. To determine the stratigraphic patterns that will be produced by the natural dynamics of these environments.
3. To determine, through simulation of a deltaic depositional system, the relationships of the resulting stratigraphic units to geologic time.

Important Terms

Alluvial plain—The upper portion of a river's depositional area, in which deposition outside of the channels takes place only during floodstage.

Backswamps—Low-lying, poorly drained areas along the channel margins, separated from the channels by natural levees. A *crevasse-splay* deposit is formed when the natural levee is breached, and a miniature delta is built out into the backswamp area.

Delta—The deposit, normally forming new land, at the mouth of a stream or river. In map view deltas are often triangular, like the Greek letter for which they are named (see Figure 14.1A).

Deltaic plain—The lower portion of a river's depositional area, near where it enters the sea. The main channels split (*bifurcate)* into many separate channels, or *distributaries.*

Distributary mouth bar—A deposit of sand and silt at the mouth of a distributary channel.

Meanders—Natural winding loops in the course of a stream channel.

Natural levees—Elevated ridges along a stream bank, caused by the deposition of more sediment nearer to the channel than farther away during floodstage.

Point bar—An accumulation of sand, silt, or gravel on the inside of a meander loop.

Introduction

More detritus is deposited in deltaic systems than in any other environmental complex. Large deltas occur at the mouths of most major river systems. Their sediments may be derived from a vast continental interior, such as those of the Mississippi and the Amazon, or they may come from a small, rapidly eroding mountain chain near a coastline. Barrier island shorelines, studied in the last chapter, are areas where sediment originally from a delta has been redistributed along an entire seacoast by longshore currents. Different delta types (Figure 14.1) are produced by differences in the amount of sediment supplied and the effects of waves and tidal currents. Rivers with less sediment produce more lobate deltas, and those with more sediment produce the elongate type of delta, of which the Mississippi River provides an important example.

A deltaic plain (Figure 14.2) is characterized by bifurcating distributaries, mouth bars, and marine waters in proximity. On an alluvial plain, however, rivers flow rapidly enough to maintain a single course, rarely overtopping their banks, and leaving the floodplain dry and unaffected by sedimentation except at floodstage. In a deltaic plain, in sharp contrast, rivers are slowed by their entrance into the sea and can no longer carry their sediment load; this causes them to bifurcate into many separate channels in an attempt to cut through their own dumped sediment.

Because sand is carried and deposited in more faster moving water than silt and clay, sandy sediment is found nearer the channels, and muddy sediments farther away. Successive flooding of a channel causes the deposition of more sediment along the channel margins, forming natural levees (Figure 14.2), which eventually stand up above the surrounding area and act as barriers between the stream and the floodplain. The poorly drained backswamps will receive silt and clay near the channels and may accumulate plant debris (to become coal) and carbonates (to become freshwater limestones) away from the channels.

The formation of point bar deposits by stream meanders is shown in Figure 14.3. Because the stream channel tends to maintain a constant cross-sectional area as it migrates, it cuts away material from the outside band (A′) as it fills on the inside bank (A). As the cross section shows, point bars are usually cross bedded at right angles to the direction of flow. The entire channel constantly migrates and may leave a series of natural levee ridges on the point bars. Eventually the stream may "cut off" the meander (Figure 14.4) and again take a more direct, straighter course. In doing so, it cuts through its own sediments again, producing an erosion surface. The abandoned meander bend now has no detrital sediment supply except at times of flooding, when it receives some silt or clay. Abandoned meanders filled with water become *oxbow lakes.* A meandering stream eventually produces a sheetlike bed of sand or gravel that is as wide as its floodplain, the lower surface of which is always an unconformity.

The formation of distributary mouth bars is shown in Figure 14.5. Distributaries do not meander, but deposit sandy sediment principally at their mouths. Eventually the pile of sand at the mouth becomes very large, and the channel must split or bifurcate to get around the bar, forming two new distributaries, which begin building their own mouth bars. In cross section (Figure 14.5), an area that starts out as a muddy bottom receives coarser and coarser sediment as the channel advances, until the sandy bar itself has arrived. A vertical sequence such as this, that goes from fine to coarse sediment, is called a *fill-in.*

Fill-ins result from the natural tendency of water to run downhill, carrying sediment with it. Sediment fills in the low spots; when one low area is filled, the channel splits and begins filling in another. This process takes place at many scales of magnitude in a delta. The smallest scale is observed in the formation of distributary mouth bars; a single distributary may be only 10 meters wide. Many such channels (and backswamps) together combine to deposit a larger mass of sediment, such as the Cubits Crevasse sequence (Figure 14.6 and area 4 in Figure 14.7A), which is a fill-in about 16 kilometers wide. The Cubits Crevasse complex is only one of eight major fill-ins that form the present birdfoot delta, a mass of sediment about 50 kilometers wide, which began forming in

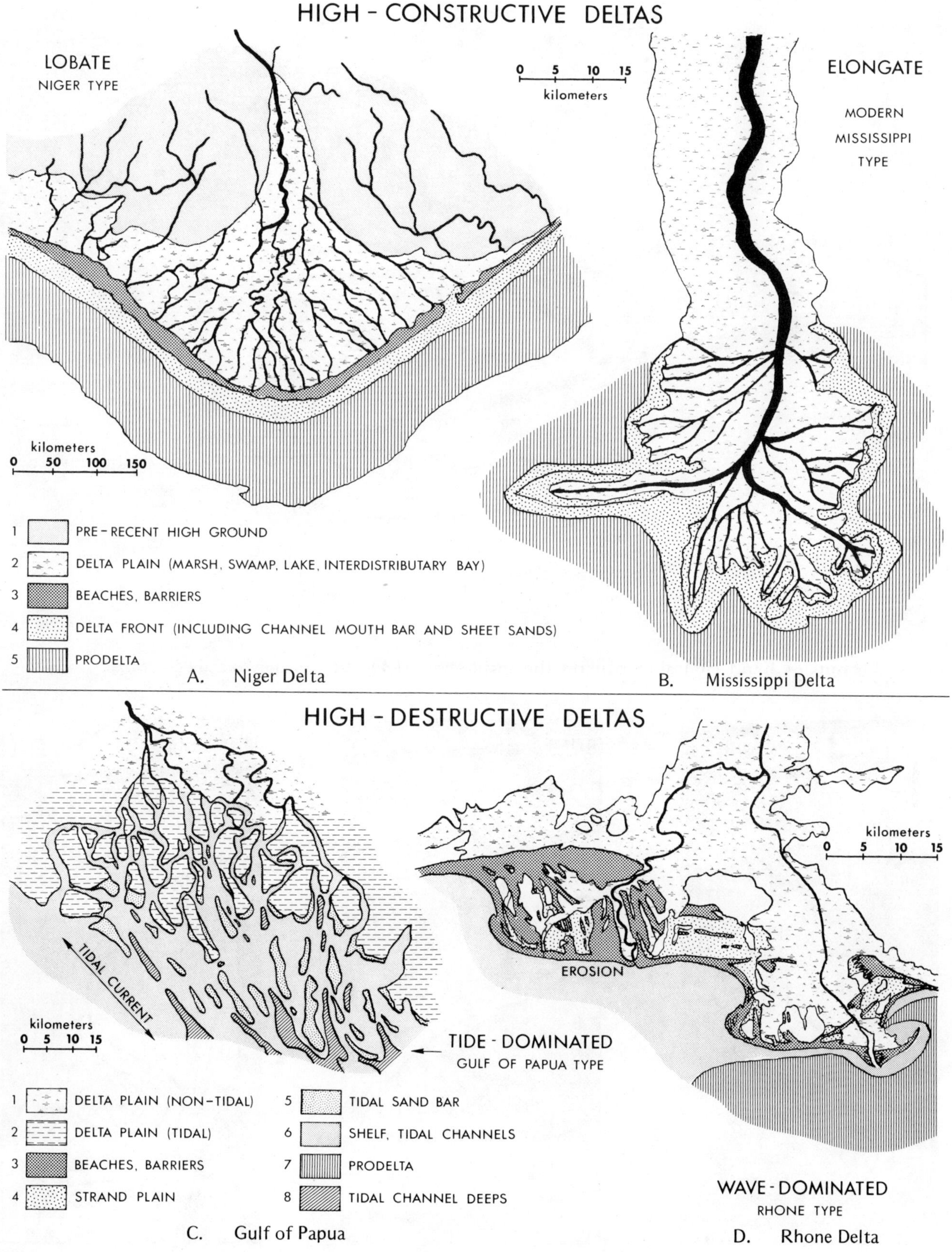

Figure 14.1. Basic delta types.

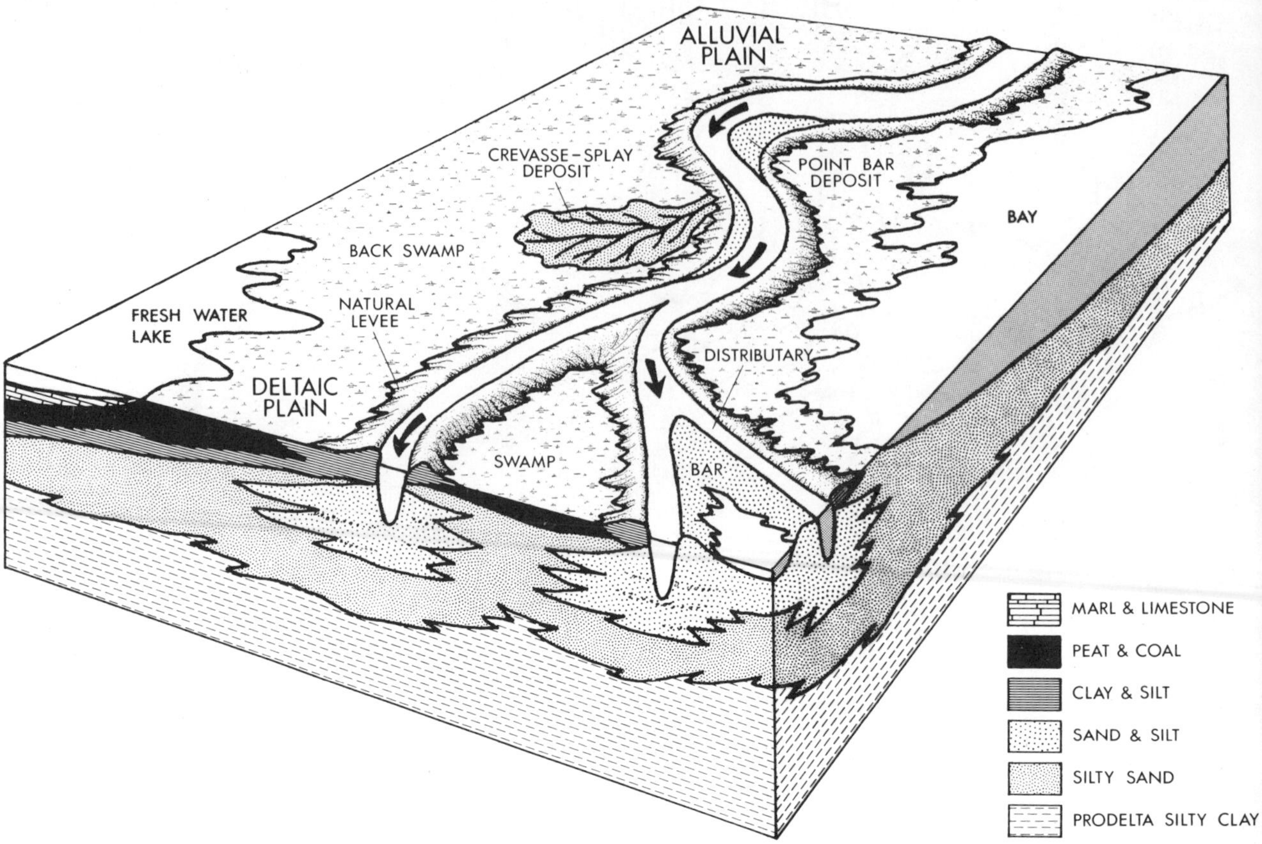

Figure 14.2. Depositional environments and their sediment types on an alluvial and deltaic plain.

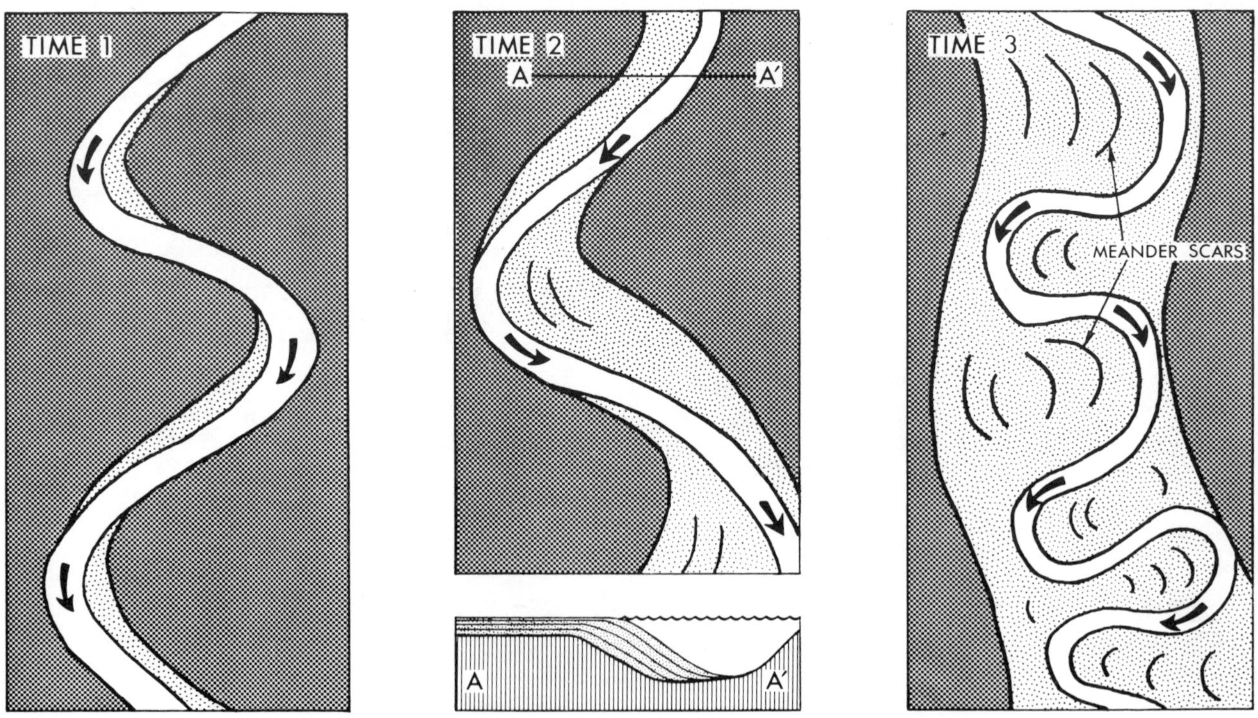

Figure 14.3. The formation of meanders and point bars.

Figure 14.4. Aerial photograph of the Mississippi River's alluvial plain in Louisiana, showing a cutoff meander loop, oxbow lake, point bars, natural levees, and floodplain areas.

the early 1800s (Figures 14.7A and 2.5). The entire Mississippi delta complex is a larger mass of sediment made up of at least seven such subunits, covering about 300 kilometers along the coast and extending about 150 kilometers away from the alluvial plain (Figure 14.7B). The development of this entire complex took place over only a few thousand years. To fill in very large areas of a continental shelf or an inland sea by this process of migrating deltaic lobes would take at least several million years. The patterns of time surfaces or isochrons in such rocks must be very complex.

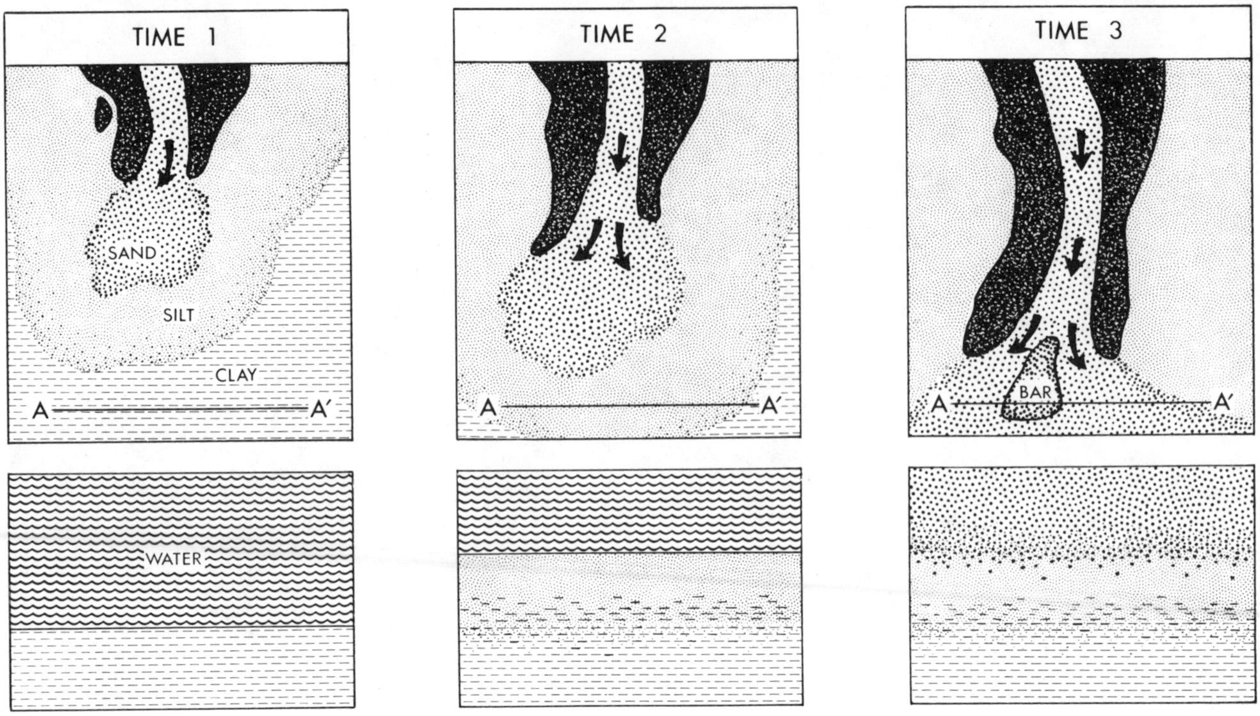

Figure 14.5. Development of a distributary mouth bar.

Figure 14.6. Sedimentary fill of Bay Rondo after the development of Cubits Crevasse in 1860; compare with area 4 in Figure 14.7A.

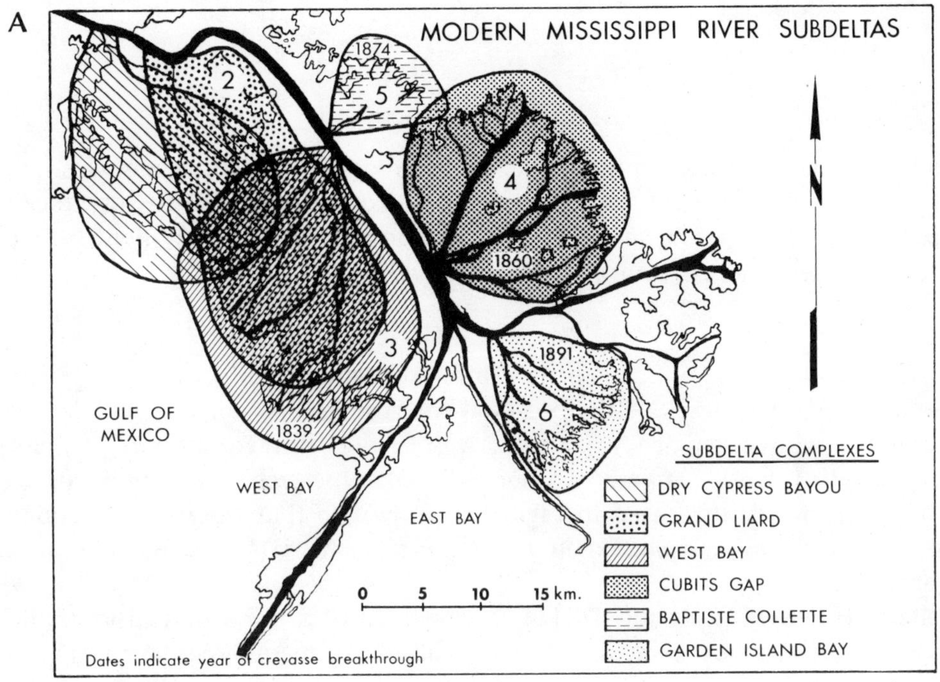

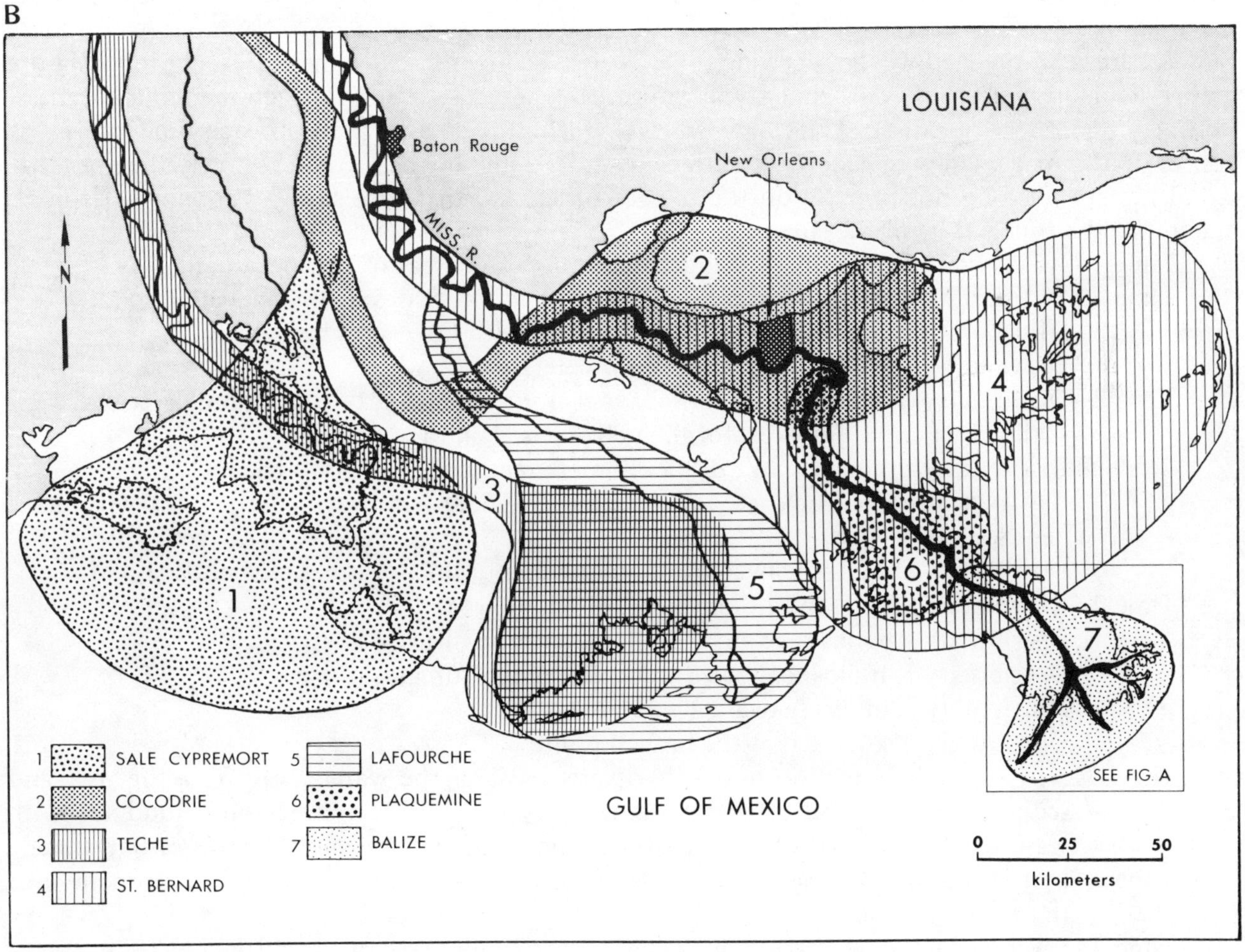

Figure 14.7. Subdeltas of the present birdfoot delta (A) and older deltaic fill-ins (B). Note that the entire area shown in A is area 7 in B.

Exercise

Part I (Optional)

Rock samples or suites of rock samples collected from deltaic and alluvial environments will be set out in the laboratory. Identify the rock type, organic content, skeletal fossils, trace fossils, and sedimentary structures in each. Using *all* of the information at your disposal, identify as precisely as possible the specific depositional environment of each.

Part II—Simulation of Deltaic Deposition

You can now "build" a three-dimensional volume of deltaic sediment, roughly similar to the Cubits Crevasse fill-in (Figure 14.6). No depositional environment remains in one place very long in a delta. If you could stand in one place (as at point A in Figure 14.8) for a hundred years or so, you could see sand, clay, plant debris, and many other sediment types accumulating in succession at the same spot. Such a sequence is illustrated in Figure 14.9. Notice that the vertical sequence of rock types is roughly the same as that shown in the traverse A-A′. Why are the lateral and vertical sequences similar?

The worksheets (Figures 14.10 and 14.11) show a map of a hypothetical lower deltaic plain. The map shows channels, swamps, bars, and their associated sediment types. A possible sequence of events for successive points in time is given below. On sheets of tracing paper redraw the sediment map, using the symbols given in Figure 14.9 (or a standardized variety of colors), for each time specified.

All areas on the map will be receiving some type of sediment at all times. By superimposing the sediments deposited at different times, you can generate a cross section through the sediment mass. You may choose the location of the A-A′ traverse; fill in the cross section for each time interval as you draw the maps. Draw in each time surface as a dark line in the cross section. Remember that sediments are piled up horizontally on top of each other, and that time surfaces are not horizontal, but follow the topography of the substrate.

Time 1—As shown in Figure 14.11. Fill in mouth bars and areas of marine silt and clay.

Time 2—Distributary A builds seaward 1 kilometer and bifurcates, forming channels F an G.
Distributary B builds seaward and combines with C to form E.
Distributary D builds seaward 0.5 kilometers.

Time 3—Distributary F builds its new channel 1 kilometer seaward; G builds seaward by 1 kilometer and bifurcates, forming channels H and I.
Distributary E builds seaward 0.5 kilometer and bifurcates into channels J and K.
Bay Y (which was enclosed when B and C joined) is now a swamp; fill it with plant debris, clay, or marl.
Distributary D builds seaward 1 kilometer and bifurcates into channels L and M.

Time 4—Distributaries F, H, and I build seaward 1 kilometer.
Distributary J builds seaward 1.5 kilometers and bifurcates.
Distributary K builds seaward 1 kilometer and bifurcates.
Distributary L builds seaward 1 kilometer.
Distributary M is closed at the branch point.

Time 5—New crevasse formation upstream diverts most of the water away from the area, and sediment compaction causes the entire area to subside and become inundated with seawater. Oyster banks form over the shallowest areas, leaving a deposit of limestone.

Time 6—Distributaries from the crevasse splay upstream have worked their way diagonally into this area from the northwest.

Time 7—Several of the new distributaries continue to build seaward and bifurcate.

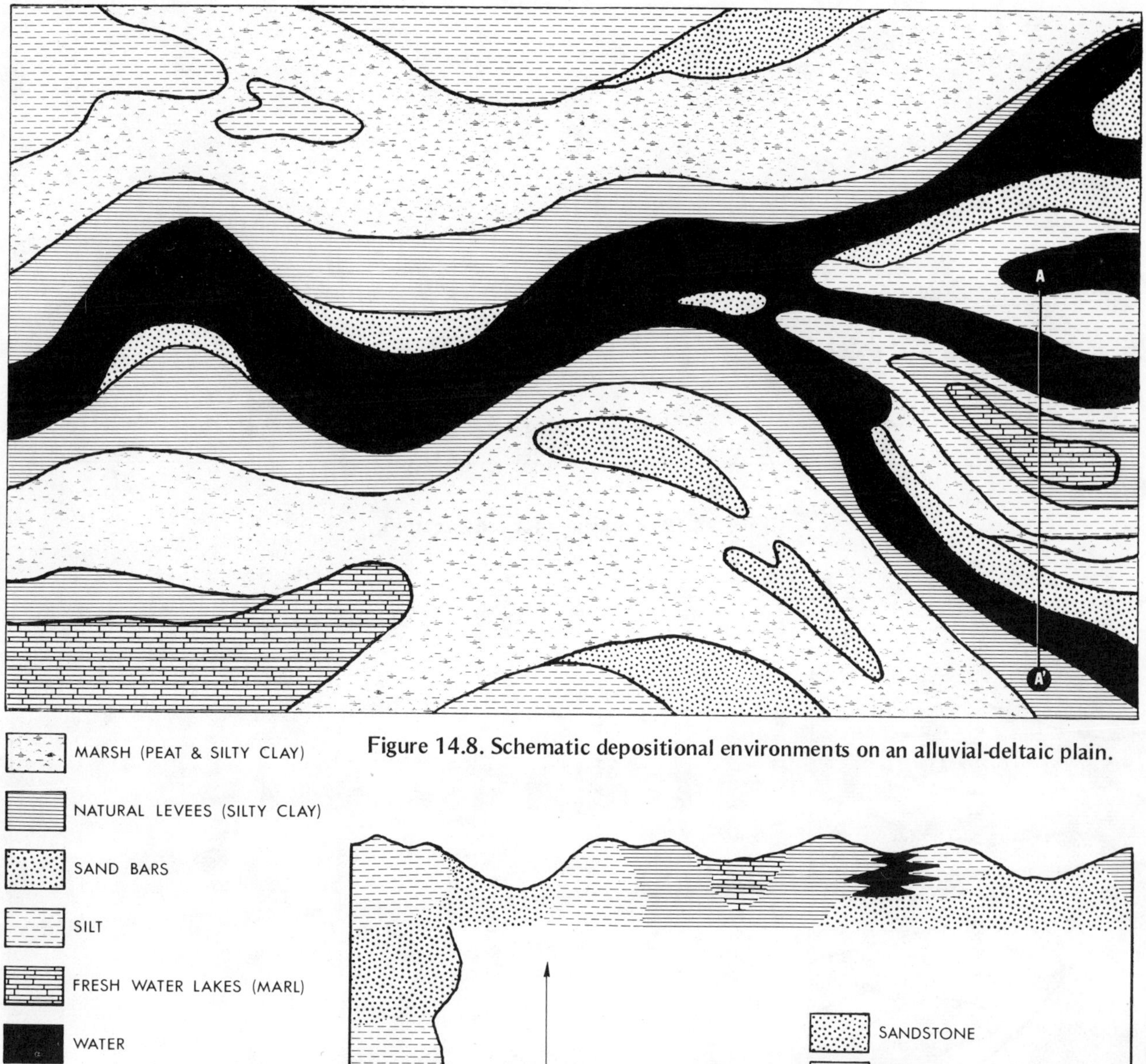

Figure 14.8. Schematic depositional environments on an alluvial-deltaic plain.

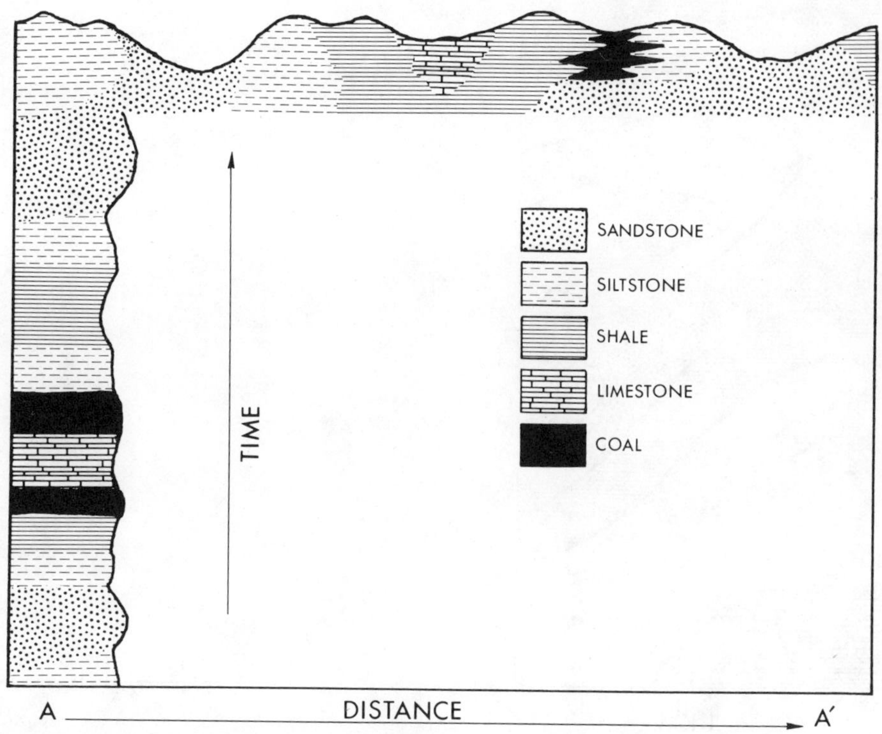

Figure 14.9. Sedimentary rock types that could be produced by the lateral sequence of environments shown along traverse A-A′ in Figure 14.8.

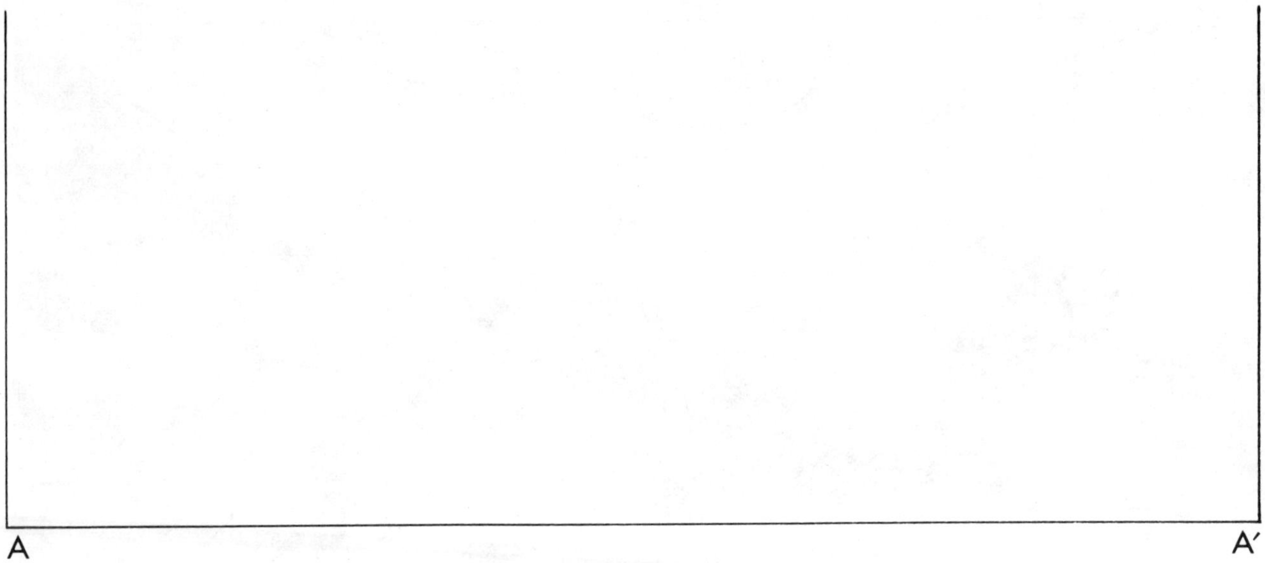

Figure 14.10. Worksheet for cross section.

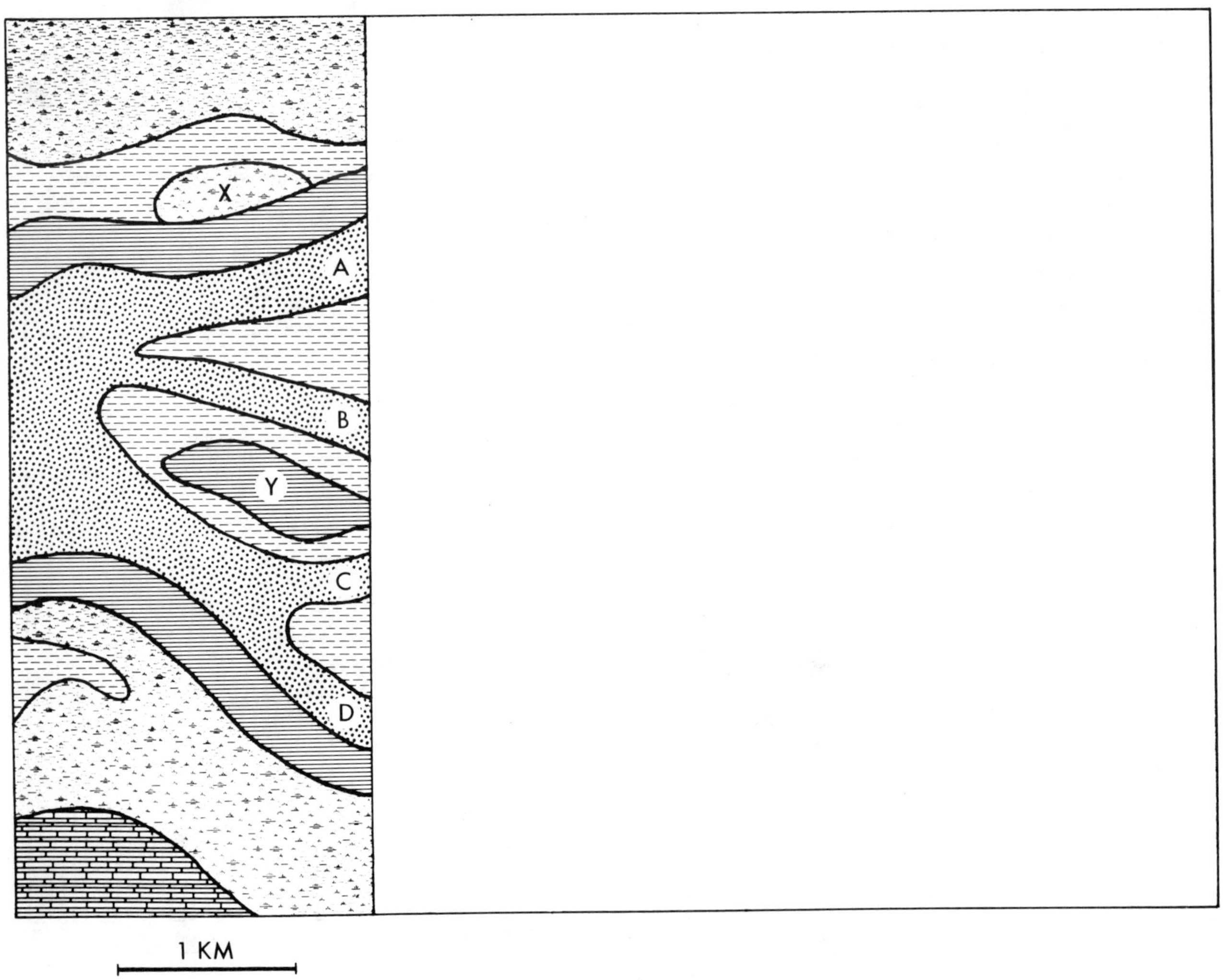

Figure 14.11. Worksheet for map view.

Questions for Discussion

1. Are there any areas in Figure 14.7 that had a history similar to the simulation? How many times have areas in the Mississippi delta system been inundated by seawater and then later experienced another series of distributary advances? How large an area was inundated each time?
2. Geologists have observed in Carboniferous-age rocks that the same sequences of coal, shale, sandstone, and limestone may repeat themselves over and over again in the inland sea deposits of the eastern and central United States. If these are deltaic deposits, what causes the apparent cyclicity? Does it mean that the sea completely advanced and retreated over the entire continent once for each cycle?
3. What is the relationship between time surfaces and homogeneous rock units in a body of sediment such as the Cubits Crevasse complex? What is the relationship between time and stratigraphy among the larger subunits of a deltaic complex?

Background Reading

Blatt, H., Middleton, G., and Murray, R. 1972. Origin of Sedimentary Rocks, pp. 197-205. Englewood Cliffs, N.J.: Prentice-Hall.

Morgan, J. P., ed. 1970. Deltaic Sedimentation, Modern and Ancient. Society of Economic Paleontologists and Mineralogists, Special Publication 15.

Chapter Fifteen

Glacial Geology Patterns of Glacial Sedimentation

Objectives

1. To examine in detail the sediments and depositional environments associated with the margins of continental ice sheets.
2. To determine the stratigraphic patterns that will be produced by the dynamic processes of continental glaciers.
3. To determine, through simulation of a glacial system, the relationship of the resulting stratigraphy to geologic time.

Important Terms

End moraine—A ridgelike accumulation of till melted out of the glacier's margins. A *terminal* moraine is the end moraine built at the terminal margin of the extent of a glacial lobe.

Glacial lobe—A tongue-shaped ice mass associated with the margins of a continental glacier, caused by pressure differences in the ice sheet as well as the topography of the preglacial surface.

Lake (lacustrine) deposits—Silt, mud, and organic matter deposited in lakes. Glacial lakes commonly formed when melting water from an ice sheet was entrapped between the ice and an end moraine. Lake silts are often *varved* (see Figure 8.1G) due to the annual turnover of the lake water. Sand dunes often formed on the beaches of glacial and postglacial lakes, on the shorelines away from the prevailing wind direction. Some lakes harbored large numbers of snails and clams that formed a mixture of detrital mud and shell debris called *marl.* Lakes that have nearly filled up with sediment become swamps in which *peat* and other plant debris accumulate.

Loess—Wind-deposited clay and silt blown off outwash plains that have not yet been stabilized by vegetation. The very cold ice sheets generate strong winds around their margins that will lift and redeposit sand, silt, and clay. Both loess and sand dunes will occur on the side of an outwash plain away from the prevailing wind direction.

Outwash—Stratified (bedded) sand and gravel deposited as point bars on braided streams flowing out of the melting ice. Outwash may be poorly sorted because it has not been extensively worked by water, and it can very much resemble till.

Till—An unstratified and poorly sorted mixture of clay, silt, sand, and pebbles deposited by a glacier. The lower portions of a till sheet may contain twigs, branches, and organic matter that were incorporated into the ice when glaciers overrode ancient forests.

Introduction

Ice sheets covered large portions of the Northern Hemisphere continents during the late Cenozoic era and still remain in Greenland and Antarctica. Widespread earlier glaciations are represented by now-lithified glacial sediments from the Permian, Ordovician, and late Precambrian.

Glaciers occur not only as continent-wide sheets of ice kilometers thick, but also as tongues of ice that flow down the valleys of many mountainous areas. Alpine glaciers have little long-term effect on the rock record because they form in rapidly eroding areas (and in fact greatly increase erosion rates where even their own sediments will not survive long. Continental glaciers, on the other hand, have produced significant quantities of preservable sedimentary rocks that record unusual episodes in earth history.

Continental glaciers, like deltas, commonly deposit sediments in local, lobate masses. Observe the lobate shape of the end moraines shown in Figure 15.1. The present shapes of Lake Michigan and Lake Erie preserve the former shapes of Pleistocene ice lobes. The earliest water in Lakes Michigan and Erie (Glacial Lakes Chicago and Maumee) was derived from the melting of the ice. The continental glacial deposits of the interior United States are a complex of highly lobate patterns.

Glacial processes consist of eroding and redistributing older soil, sediments, and bedrock by scouring off the areas over which the ice moves and transporting the sediment by ice flow to the leading edges of the ice sheet, where melting and deposition take place (Figures 15.2 and 15.3). Glacial lobes advance due to pressure in the ice mass caused by the weight of the ice "upstream." When the pressure has been alleviated by forward flow, the advance ceases. If ice no longer feeds an advancing lobe, it begins to melt and retreats by melting. As the ice melts and retreats, the sediment incorporated into it is released and deposited as till, outwash, loess, or lake deposits. Both ice lobes and outwash plains can bury preglacial grasslands and forests. The winds created by the meteorological conditions along an ice front will rework the loose outwash (not yet stabilized by plant cover) into sand dunes and loess.

Ice, like a very viscous liquid, generally flows downhill. Consequently, once a low-lying area has been filled in by sediments, newly formed lobes advancing from the glacier will seek out other areas. A lobe can also advance over the same area several times (Figure 15.4) until it has been aggraded enough to cause new lobes to flow elsewhere. Thus glacial sedimentation is somewhat analogous to the *fill-ins* of deltaic deposition.

Figure 15.1. Glacial geology of parts of Michigan, Ohio, and Indiana.
Blue—Lake sediments
Yellow—Outwash
Orange—Ice-contact stratified sediments
Dark Green—Terminal moraines
Light Green—Till plains
Pink—Older till plains
Gray—Unglaciated areas
Blue lines—Striation grooves
Red lines—Beach deposits

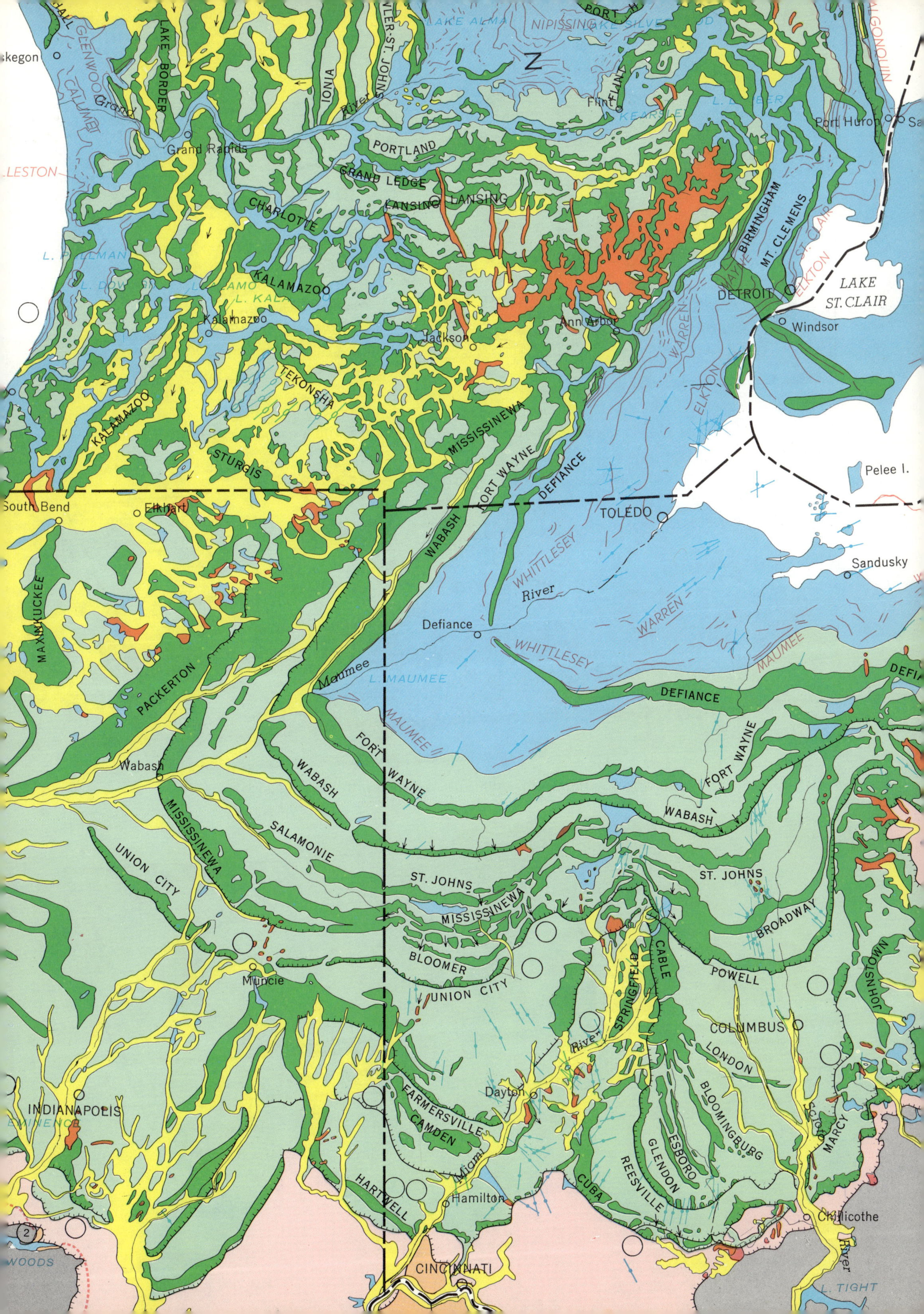

LAKE BORDER
IONIA
NIPISSING
FLINT
Flint
KEARSLEY
Port Huron
ALGONQUIN
GLENWOOD
CALUMET
Grand
River
Grand Rapids
LESTON
PORTLAND
GRAND LEDGE
LANSING
LANSING
CHARLOTTE
KALAMAZOO
Kalamazoo
Jackson
Ann Arbor
BIRMINGHAM
MT. CLEMENS
ST. CLAIR
ELKTON
DETROIT
LAKE ST. CLAIR
Windsor
WARREN
ELKTON
TEKONSHA
KALAMAZOO
STURGIS
MISSISSINEWA
FORT WAYNE
DEFIANCE
Pelee I.
South Bend
Elkhart
TOLEDO
WABASH
WHITTLESEY
River
Sandusky
WARREN
MAXINKUCKEE
Defiance
WHITTLESEY
MAUMEE
Maumee
L. MAUMEE
PACKERTON
DEFIANCE
MAUMEE
FORT WAYNE
Wabash
WABASH
FORT WAYNE
WAYNE
WABASH
MISSISSINEWA
SALAMONIE
UNION CITY
ST. JOHNS
MISSISSINEWA
ST. JOHNS
BROADWAY
CABLE
SPRINGFIELD
BLOOMER
POWELL
JOHNSTOWN
Muncie
UNION CITY
COLUMBUS
River
LONDON
Dayton
FARMERSVILLE
CAMDEN
INDIANAPOLIS
EMINENCE
BLOOMINGBURG
ESBORO
GLENDON
Scioto
MARCY
HARTWELL
Miami
Hamilton
CUBA
REESVILLE
Chillicothe
River
WOODS
CINCINNATI
L. TIGHT

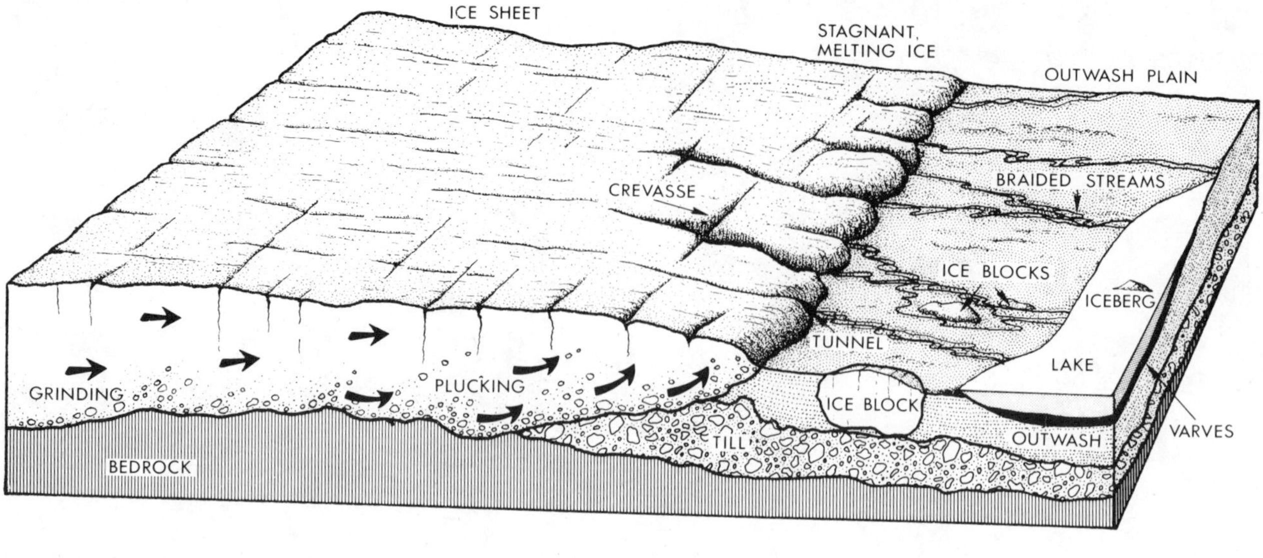

Figure 15.2. Physical processes and sediments associated with a stagnant ice front.

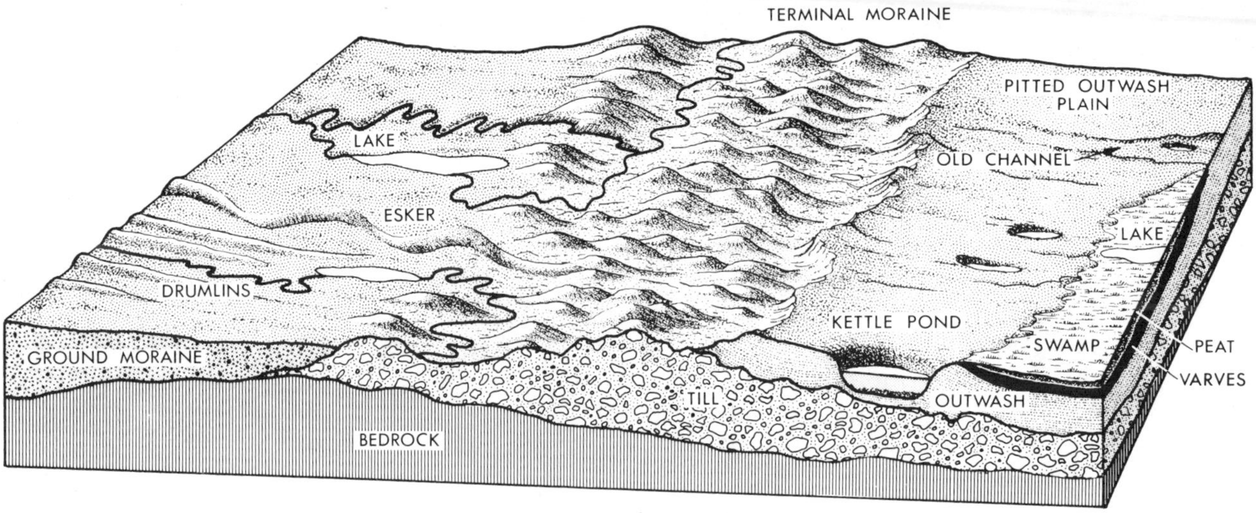

Figure 15.3. Landforms and sediments associated with a retreated ice front.

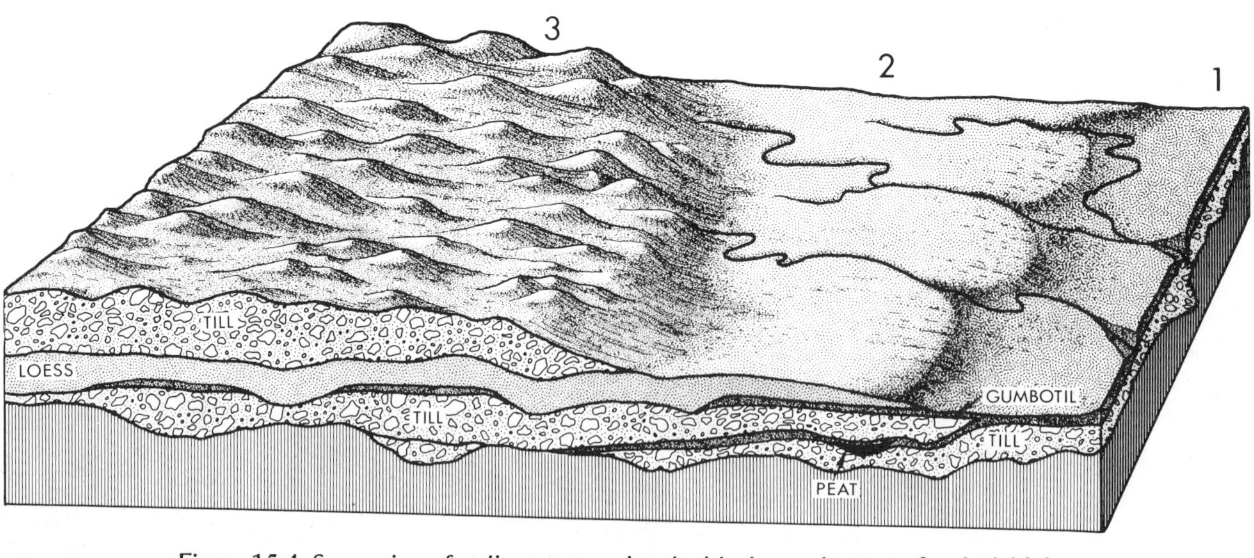

Figure 15.4. Succession of sediments associated with three advances of a glacial lobe.

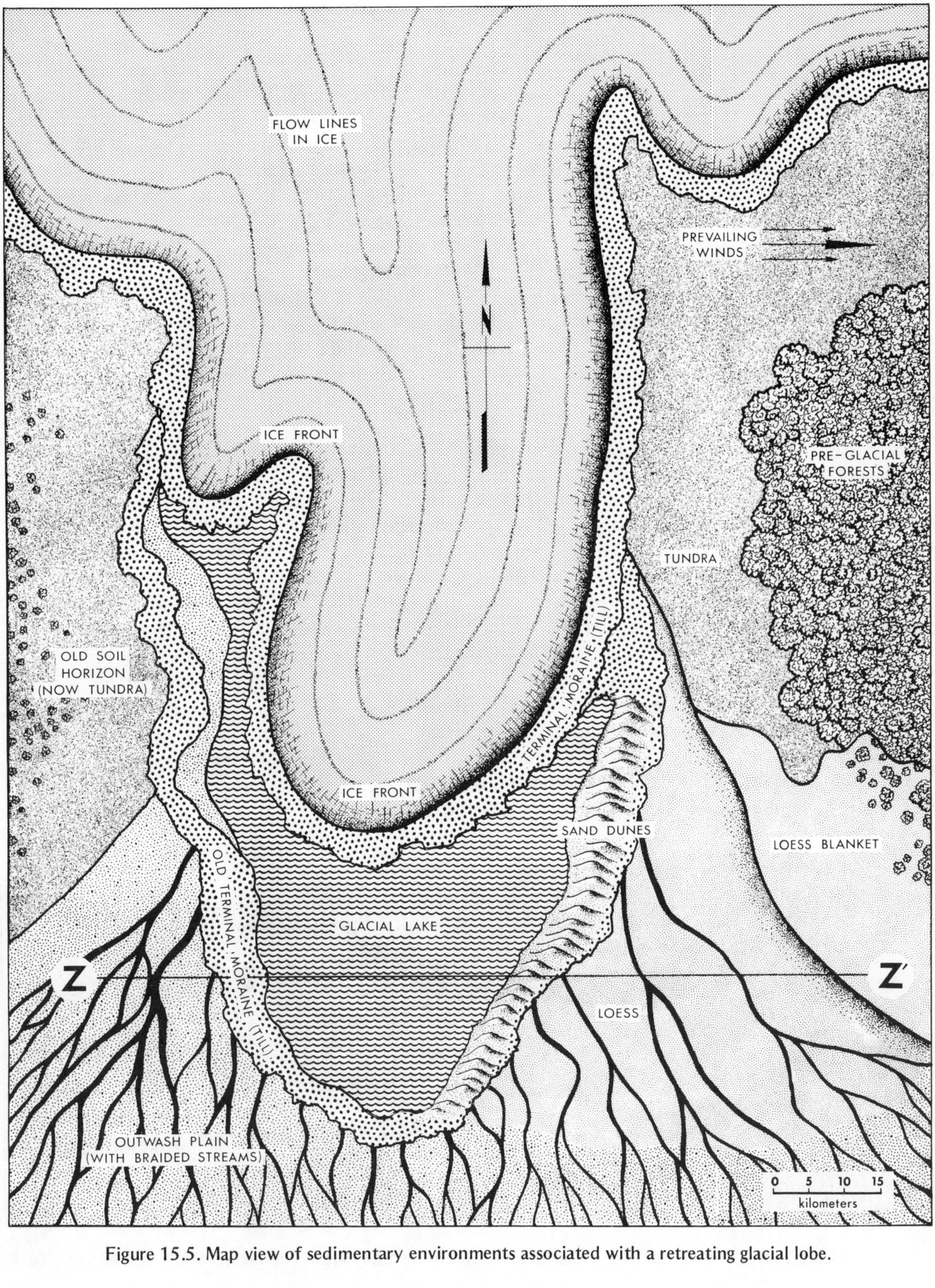

Figure 15.5. Map view of sedimentary environments associated with a retreating glacial lobe.

Exercise

Part I—Glacial Sediments and Rock Types

Sediment samples or sample suites collected from different glacial environments will be set out in the laboratory. If they are available, ancient lithified glacial materials will also be provided. Identify the sediment type, organic content, and any fossils (plant or animal) in each known and unknown. Identify the most likely source of each sample in Figure 15.5.

If time permits, examine the sorting of one or more samples of glacial till and outwash by washing the sediment through a stack of nested sieves. Concentrate the sediment in a corner of the sieve and then wash it into a beaker, using only a minimal amount of water. The beakers may be dried either in the air or in a drying oven. Weigh each fraction, and present the results as a histogram. Compare these sediments to the sediments studied in Chapter 1.

Examine only the sand fraction of the glacial till, and determine the roundness of the grains (refer to Figure 1.1). Compare this sand with the sand dune material and the outwash. What different effects do wind, ice, and water have on grain roundness?

Part II—Simulation of Glacial Deposition

Figure 15.5 shows a single lobe of a hypothetical continental glacier. Figure 15.6 illustrates a cross section of the sediment that would be found along the traverse Z-Z′. Figure 15.7 shows the ice front (in map view) of a continental glacier at time 1. Lobes will advance and retreat from this glacier at particular times and deposit all of the sediments shown in Figures 15.5 and 15.6. Your assignment is to simulate this glaciation according to the following steps, and draw cross sections through the pile of accumulated sediment along the traverses X-X′ and Y-Y′. All time surfaces and sediment types must be shown in the cross sections (Figure 15.8). Use a dark line for each time surface, and standardized colors or patterns for the sediment types. Draw sediment maps for times 1 through 6, using the tracing paper for overlay sheets.

Time 2—Lobe C advances 300 kilometers south.
Lobe D advances 200 kilometers southeast.
Lobe A retreats 100 kilometers northeast.

Time 3—Lobe C advances 100 kilometers south.
Lobe D advances 100 kilometers southeast.
Lobe A advances 300 kilometers southwest.

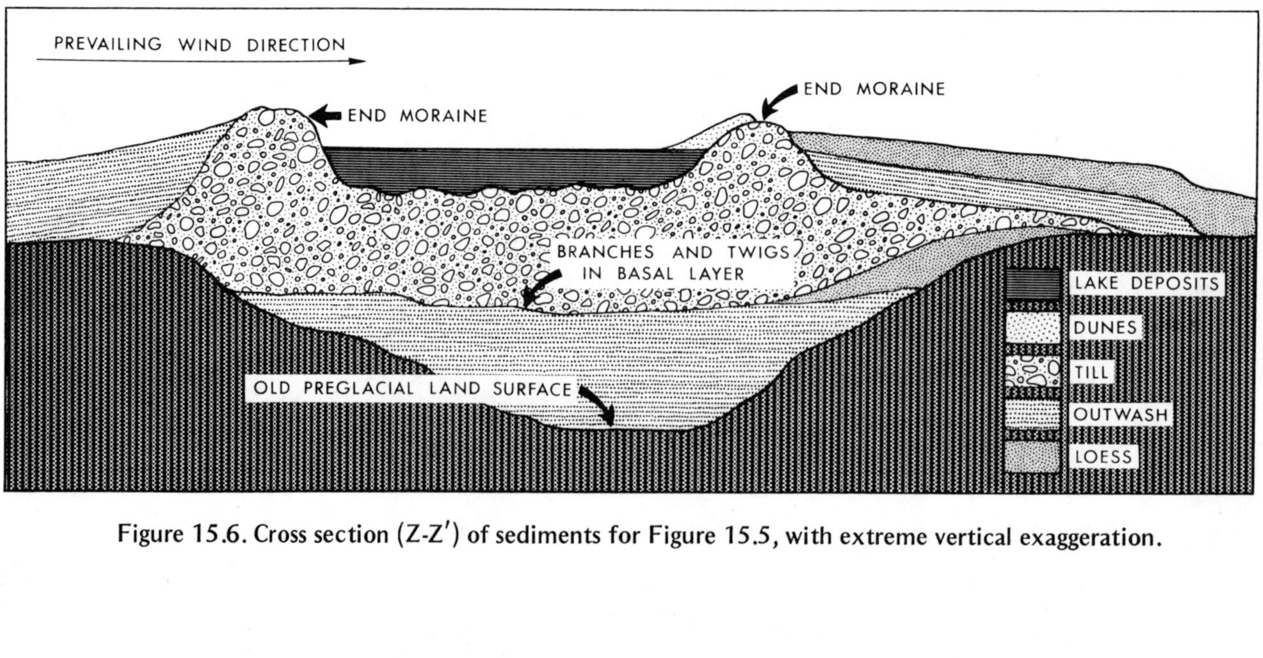

Figure 15.6. Cross section (Z-Z′) of sediments for Figure 15.5, with extreme vertical exaggeration.

Lobe B advances 100 kilometers south.
Time 4—Lobe C retreats 200 kilometers north.
Lobe A advances 200 kilometers southwest.
Lobe B advances 100 kilometers south.
Time 5—Lobe C retreats 300 kilometers north.
Lobe D retreats 200 kilometers northwest.
Lobe B advances 100 kilometers west.
Time 6—All ice retreats from area.

Questions for Discussion

1. In Chapter 14 you developed ideas concerning the relationship between time and stratigraphy in deltaic complexes. Compare or contrast these ideas with those you have developed about glaciers.
2. Are ancient glacial sedimentary rocks uniquely identifiable as such? What other depositional environments have similar lithologies? What criteria might be unique to glacial sediments?

Background Reading

Flint, R. G. 1973. The Earth and Its History, pp. 307-51. New York: W. W. Norton.

Hamilton, W., and Krinsley, D. 1967. "Upper Paleozoic Glacial Deposits of South Africa and Southern Australia." Geological Society of America, Bulletin 78:783-800.

Strahler, A. N. 1972. Planet Earth: Its Physical Systems through Geologic Time, pp. 204-9. New York: Harper and Row.

Chapter Sixteen Biostratigraphy
Patterns of Speciation

Objectives

1. To examine the environmental influences, especially geographic ones, on the origin of species.
2. To examine the geographic patterns of species distributions.
3. To determine, through simulation of evolutionary events, how the distribution of species, preserved as skeletal fossils in ancient rocks, is related to geologic time.

Important Terms

Allopatric speciation—The rise of new species from peripheral isolates, resulting in a geographic pattern of species inhabiting separate but contiguous areas (*allopatry).*

Habitat—The area (or volume) that a species occupies.

Peripheral isolate—A population on the margin of a species' range, which is reproductively isolated from the parent species and under slightly different environmental restraints.

Population—The members of a species found in some easily defined segment of the total area occupied by the species.

Species—A group of genetically similar organisms occupying a continuous range and maintaining a high level of functional similarity through both interbreeding and restraints imposed by their environment.

Sympatric speciation—The rise of a new species from populations *within* the range of the parent. The isolating mechanisms usually are related to changes in behavior that create different niches within the same geographic range. Secondary *sympatry* can result from the recolonization of the parental range by an allopatric descendant.

Introduction

Among living organisms, the most common patterns of species distributions, within, say, a genus, are allopatric. For example, seven species of the *Alectoris* partridges occupy a series of separate but contiguous geographic areas extending from the eastern Atlantic Ocean across Eurasia and northern Africa to the Yellow Sea (Figure 16.1). In only two small areas, less than 1 percent of the generic range, do the species overlap so as to be sympatric. Sympatry is generally uncommon, and requires special isolating mechanisms to provide separate ecological niches within the shared area. Otherwise, competition would drive one of the species to local extinction and restore complete allopatry. If environmental conditions change within a species' range, it is unlikely that the species will "evolve" through natural selection to adapt to the change. Rather, the range boundaries will migrate, and the species will become extinct in some areas and colonize some new ones. If migration to new areas is prevented by an impenetrable barrier, then the species could become extinct. Some of its peripheral isolates, however, could give rise to new species adapted to the new conditions.

Because of natural selection, species are environmentally adapted to the conditions in their habitat. Towards the periphery of their range they are less well adapted, and at the range boundary abundance drops to zero. The sparse peripheral populations of a species are subject to the greatest ecological stresses and are pretty much reproductively isolated. Therefore the peripheral isolates are the most likely populations to produce new species. For example, the tree snake *Dendrophis pictus* of southeast Asia has four subspecies located in small patches that occur along the margins of the parental range (Figure 16.2). The four subspecies are, in all probability, the beginnings of new allopatric species of this group of animals.

Consider the complex mosaic of environmental conditions that would be found in the benthic habitats along a broad continental shelf or an expansive epicontinental sea. Habitat differences could be caused by nutrient supply, substrate differences, salinity, light penetration, depth, temperature, and wave and current activity. It is probable that these habitat differences would be geographically distributed on a seafloor and strongly influenced by the location and configuration of the shoreline, especially by the locations of river mouths that supply large quantities of detrital sediment. Thus benthic marine species are just as likely to be allopatric as any others.

Exercise—Simulation of Allopatric Speciation

The base map for this exercise is Figure 16.3, which shows a portion of a marine shelf. The area is inhabited by nine species of a single genus of some generalized benthic marine invertebrate. Species A lives on the detrital sediments immediately offshore of a large marine delta. Species B and C inhabit sandy shoreline environments on either side of the delta. Species D and E are found on silty mud bottoms of the nearshore shelf. Species F, G, and H inhabit geographically separated carbonate banks. Species I lives in the mixed detrital clay and skeletal carbonates of the outer shelf. All of the species are completely allopatric.

Part I

The first part of the simulation involves speciation alone, without considering sedimentation. Speciation events will be represented as occuring at times 1 through 5. Time 1 is represented by the nine species on the base map. For time 2 and each subsequent time, determine what happens to each species by tossing a coin. Heads indicates that the species will produce at least one allopatric species somewhere along its margins (choose a location *randomly,* by tossing coins or throwing a pencil down on the paper), and tails indicates it will not produce any new species. Label all new species alphabetically, starting with J. Each new species will compete with preexisting species in their habitats. For example, if F gave rise to J on its southwest side, J would be in competition with D. For each case of this sort toss the coin again: heads, the older species survives and pushes out the new one, and tails, vice-versa. Do not subdivide any of the habitats into smaller areas, and do not assume any sympatry. For each successive time interval, each species, including all new ones, will have an equal probability of speciating or not. A new species will not have this chance until the next round. Remember that for each time interval there will always be nine species in the area.

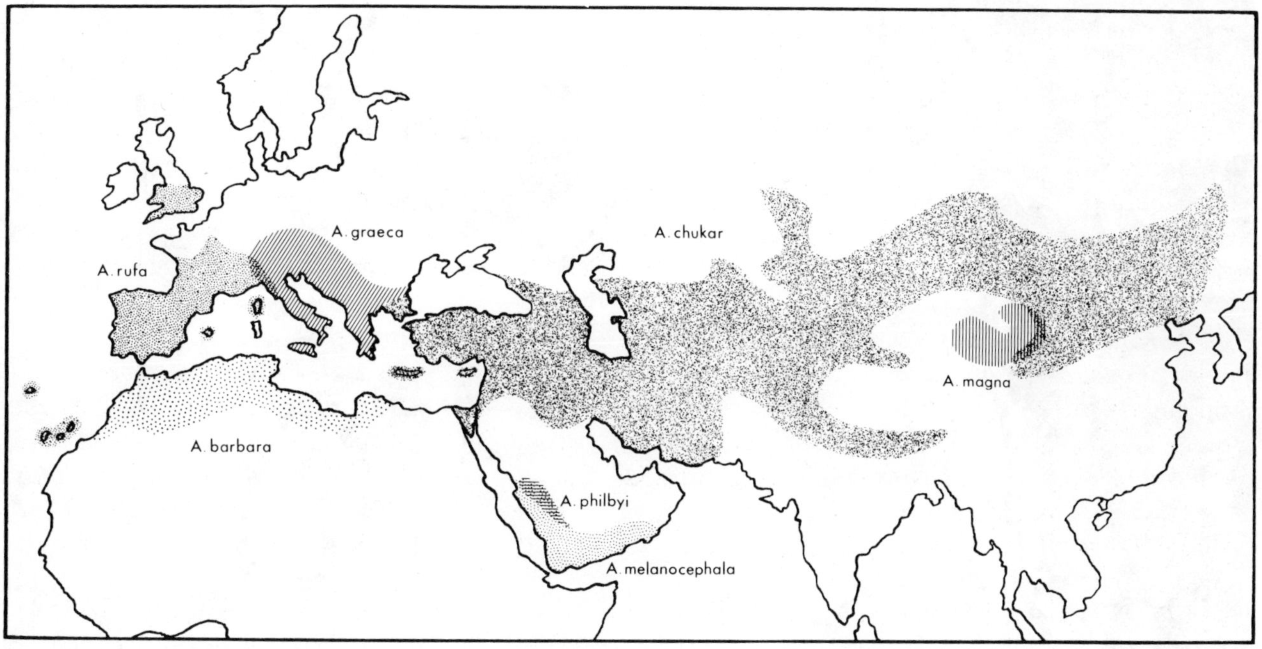

Figure 16.1. Pattern of geographic speciation in the *Alectoris* partridges.

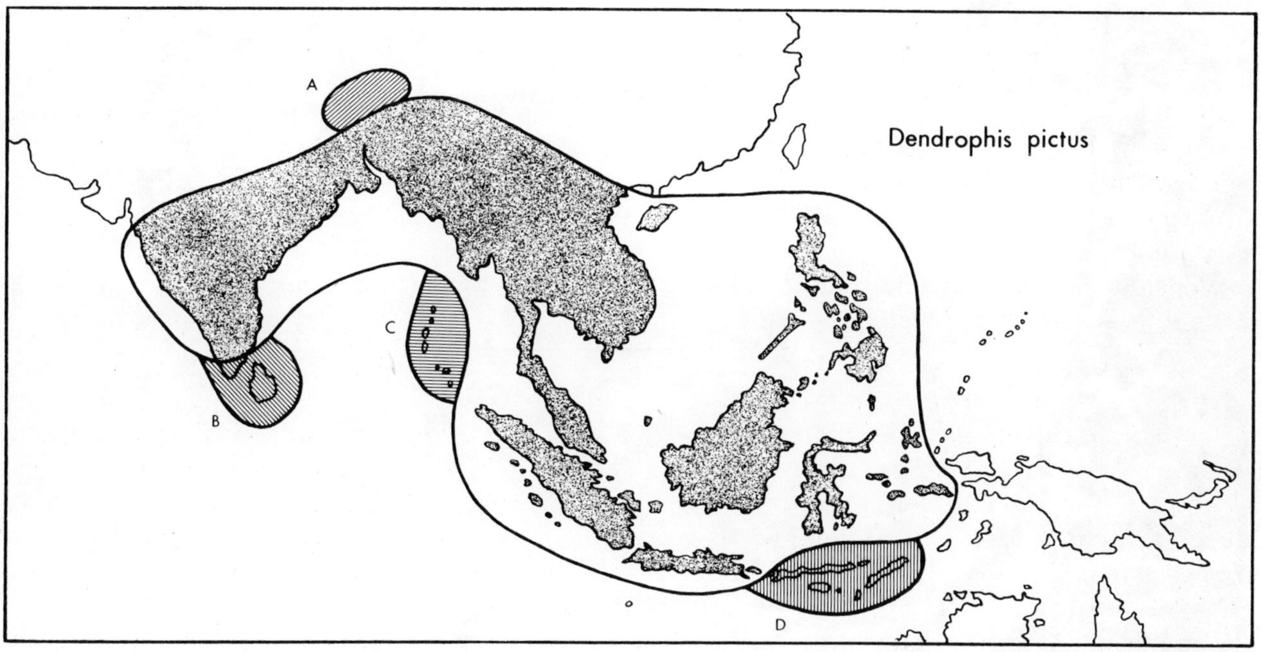

Figure 16.2. Range of the tree snake *Dendrophis pictus,* and peripherally developed ranges of four subspecies.

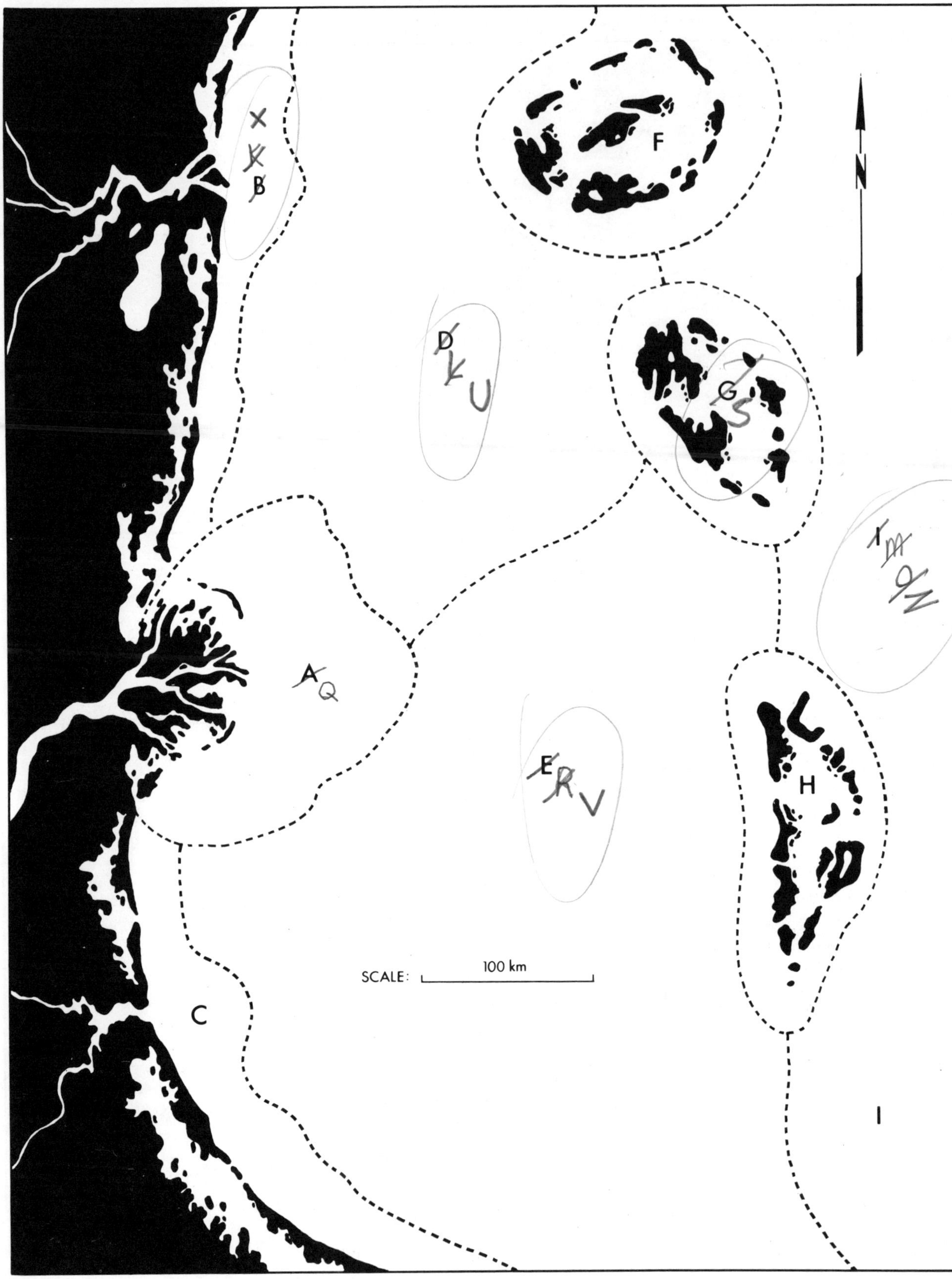

Figure 16.3. Base map for exercise.

Keep track of what happens to each species by showing their lineages as a family tree similar to Figure 3.1. Use horizontal lines, numbered from the bottom up, for times 1 through 5. Use vertical lines for each species, with side branches to denote daughter species placed to represent their times of origin. Keep track of their map locations by crossing out the letter and adding a new one when a species is replaced. In a trial run, by time 5, only 3 of the original 9 species survived, and 14 allopatric species were produced, 7 of which successfully displaced established competitors. The chart you have made showing what happens to each species through time is called a *phylogenetic tree.*

If time permits, you may also want to simulate morphology by coin tossing. Assume some simple geometric shape for each organism, and, using your imagination, list 5 to 10 features that can be described as present (heads) or absent (tails). For example, you might choose a simple circle as the basic shape. One feature could be having or not having a tail, shown as a simple line from the 6 o'clock position. Other features could be heads, appendages, cross-bars, and so on. Each time a *successful* new species arises, retossing the coin for each morphologic feature will provide a randomized simulation of morphological change.

Part II

The second part of the simulation will involve deposition of sediment and burial of the species as skeletal fossils. Use the overlay sheets, as in previous exercises, and map the distribution of species (as simulated in the first part of the exercise) as incorporated into the rocks according to the following pattern.

Time 1—Environments as shown in Figure 16.3.

Time 2—Shoreline regresses by 200 kilometers.

Time 3—Shoreline regresses by 200 kilometers more.

Time 4—Shoreline transgresses by 300 kilometers.

Time 5—Shoreline transgresses by 200 kilometers more.

Construct a stratigraphic column that represents the vertical thickness of rock deposited at the five localities (on the base map) that have had the largest combination of extinctions and originations. Draw each column about 2 centimeters wide, using standardized symbols or colors for rock types. Represent any surfaces or nondeposition by a wavy line. On the left of each column, indicate by brackets the time intervals represented in each part of the section. On the right of each column, indicate by brackets the species that are present as skeletal fossils in each part of the section.

How well does the vertical sequence of species at each locality represent the phylogenetic tree? Which localities show a true phylogenetic sequence? Which localities have a parent species overlying a daughter (a false phylogeny)? Which localities show displacement of one lineage by another? Which localities show stratigraphically recurrent species?

Questions for Discussion

1. If you wished to trace through geologic time a series of populations that inhabited an offshore carbonate facies of a marine shelf sequence, would it be more useful to study the vertical sequence of fossils at one locality or the lateral sequence of fossil types at many localities? Which plan yields a longer interval of geologic time?
2. What general relationship exists between the distribution of allopatric species and time surfaces? Is the first appearance of a species in a local rock column a good indicator of a geologic time surface?
3. How useful is the principle of superposition when used alone to determine evolutionary relationships?

Background Reading

Eldredge, N., and Gould, S. J. 1972. "Punctuated Equilibria: An Alternative to Phyletic Gradualism." In T. J. Schopf, ed., Models in Paleobiology, pp. 82-115. San Francisco: Freeman, Cooper.
Smith, R. L. 1966. Ecology and Field Biology, pp. 448-84. New York: Harper and Row.

Chapter Seventeen **Geotectonics** Sedimentation and Plate Tectonics

Objectives

1. To examine the large-scale patterns of sedimentation in relationship to plate tectonics.
2. To compare the overall stratigraphic patterns on continental margins with those in continental interiors.
3. To compare the stratigraphy of a simulated continent and its tectonic systems to an actual geologic history.

Important Terms

Graben—An elongate sedimentary basin formed by a downdropped central block bounded on both sides by high-angle faults.

Plate—A rigid polygonal segment of the earth's crust, which moves as a unit over a more mobile layer beneath the lithosphere.

Plate tectonics—The global system of crustal dynamics, involving the origin of new crust in *spreading zones,* the loss of old crust to the mantle in *subduction zones,* and the resulting movements of the continents as parts of lithospheric plates.

Seafloor spreading—The enlargement of oceanic basins by the production of new crust in a spreading zone.

Introduction

Plate tectonics provides a global frame of reference in which to understand historical geology. The movements of the crustal plates have affected the entire rock record, as well as the evolution of living things. This chapter attempts to examine how plate tectonics have influenced patterns of sedimentation over rather large areas of the earth. You should review the basic concepts of plate tectonics before beginning this exercise, either from Chapters 8 and 9 in Dott and Batten (1976) or from a text suggested by your instructor.

Each of the several stages of plate tectonic activity produces distinctive patterns of sedimentation. The stages need not occur in strict succession; in fact, they may occur simultaneously, as is true of some plates and parts of plates on the earth today. The stages are as follows.

1. Continental rifting; or splitting, leading to the opening up of new, initially very narrow, ocean basins.
2. Continental separation and high rates of seafloor spreading, associated with raised sea levels and extensive epicontinental seas.
3. Continental collision, preceded by lowered sea levels and leading to the complete closure of preexisting ocean basins. The large-scale patterns of sedimentation produced by such a series of events is observable not only along the continental margins but even in the stable interior regions of continents.

Rifting

The initial stage of continental separation is the development of a rift valley, or graben. The first sediments that normally fill in the rift are those of alluvial fans, which develop along the fault escarpments (see Figure 2.2) as a series of *fill-ins* as streams shift their courses and eventually aggrade all the low-lying areas. Alluvial fans may develop along any escarpment, tectonic or otherwise, where a sediment-carrying stream experiences a sharp change in gradient (a *nickpoint)* and deposits a large mound of poorly sorted, poorly stratified detrital sediment. Alluvial fans develop in several stages (Figure 17.1). First, upward displacement along the fault produces a steep escarpment; streams flowing across the fault encounter a nickpoint and dump their sediment load. The earliest sediments are the coarsest as well as the most poorly sorted. As deposition proceeds, the sediment mass becomes higher and wider, and the source terrain is lowered through erosion, gradually obliterating the nickpoint. The sediments become finer and better sorted upward in the fan, as well as from the fanhead to the base. In the final stage, all of the laterally adjacent fans along the scarp coalesce to form a *bajada,* a cover of sediment that has completely filled in the basin. Once the basin has been partially or completely filled, renewed movement along the fault could start the whole process over again, with a new escarpment and new nickpoints, and generate another bajada sequence on top of the old one. Towards the center of the graben or rift valley, the sediments are more like normal stream point bars, floodplains, and lake deposits.

Often, rifting underneath a continent appears to be initiated at a hot spot, where a plume of hot material is thought to be rising from the mantle. Commonly, three or so fractures radiate from the center of heating, the way mud cracks when it is drying (see Figure 8.7). The three radiating rifts form a *triple junction.* As the process continues, two of the cracks open up to form a new ocean basin. The third arm, which is now nearly at right angles to the new shoreline, becomes a sediment-filled graben termed an *aulacogen* (Figure 17.2). Aulacogenic sediments are normally intruded by dikes and sills and interbedded with lava flows coming from the hot materials welling up from the mantle. Triple junctions are not connected exclusively with hot mantle plumes; they can provide an efficient system of dissipating any mantle-derived heat, similar to the way basalt cracks into hexagonal columns (with triple junctions) as it cools. A possible sequence of sedimentary events associated with continental rifting is shown in Figure 17.3.

Epicontinental Seas

The newly formed narrow ocean basins are shallow in comparison to the abyssal plains of the very large older oceanic areas. In large oceans with spreading zones, such as the Atlantic, the water is less deep over the midocean ridge than away from it, because the oceanic crust is locally bulged up where hot materials from the mantle are rising to form new crust. As rifting proceeds, older areas of

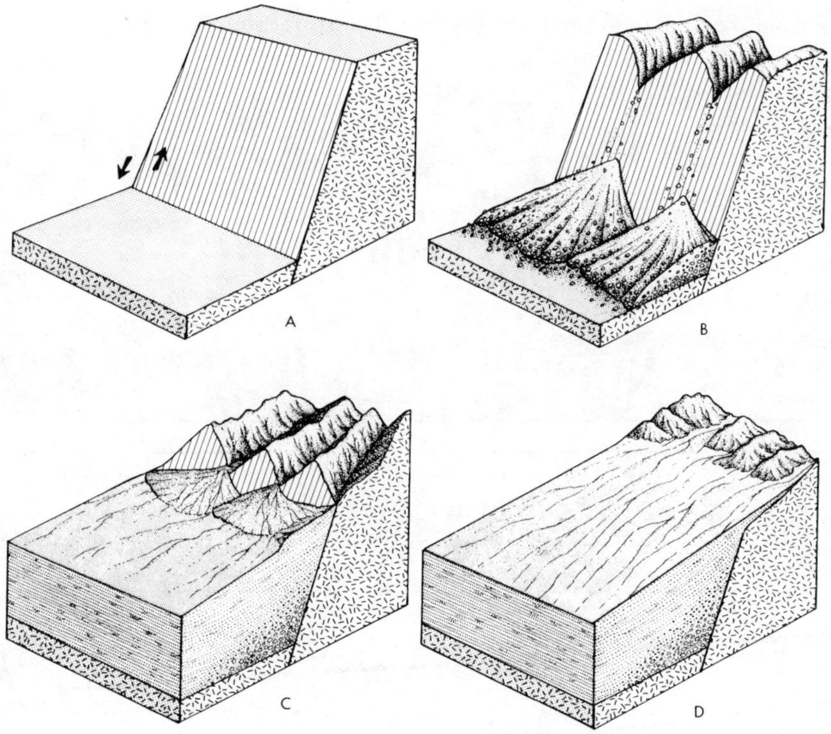

Figure 17.1. Stages in the development of alluvial fans.

deep ocean basins are gradually replaced by newly formed, upbulged, shallow oceans. This produces a gradual rise in worldwide sea levels, and continental areas become slowly covered with relatively superficial skins of seawater, called *epicontinental seas.* Hudson's Bay in Canada is a very small epicontinental sea that remains today. Most past intervals of extensive continental inundation represent times of active spreading, when relatively new, medium-sized oceans were in their most vigorous period of growth. Early stages in rifting and stages of continental collision are normally times of slow spreading.

Epicontinental sedimentation is of considerable importance in geology because it has produced most of the sedimentary rocks that we see on our present land areas, including the general mixture of carbonate and detrital rocks that you have so far studied in this manual. The major environments in such seas were deltas, which provided sediment from uplifted areas on the continental margins, shoreline sediments, carbonate banks, and open marine shelf areas. Figure 17.4 shows the deposits of four large deltas that were deposited by rivers emerging from a continental borderland and built back into the shallow seas covering the continent. The shelf sediments around a modern delta (which, incidentally, does not feed into an epicontinental sea) show specific areas of contemporaneous shoreline deposits, carbonate shelf areas away from the river mouths, and open marine shelf areas (Figure 17.5). Each of the Paleozoic deltas shown in Figure 17.4 was flanked by a similar assemblage of shoreline sands, carbonate banks, and open shelf muds or fine-grained carbonates. Figure 17.4 also ties together the environmental complexes that you have worked with in Chapters 7, 13, and 14.

Continental Margins

A continental margin may, depending on the direction of movement, be either the leading side or the trailing edge of the continent, and its history may change if the continent changes its direction of movement. The leading edge of a continent is geologically very active and generally is at or near a subduction zone, complete with volcanic island arcs and oceanic trenches, like those in the

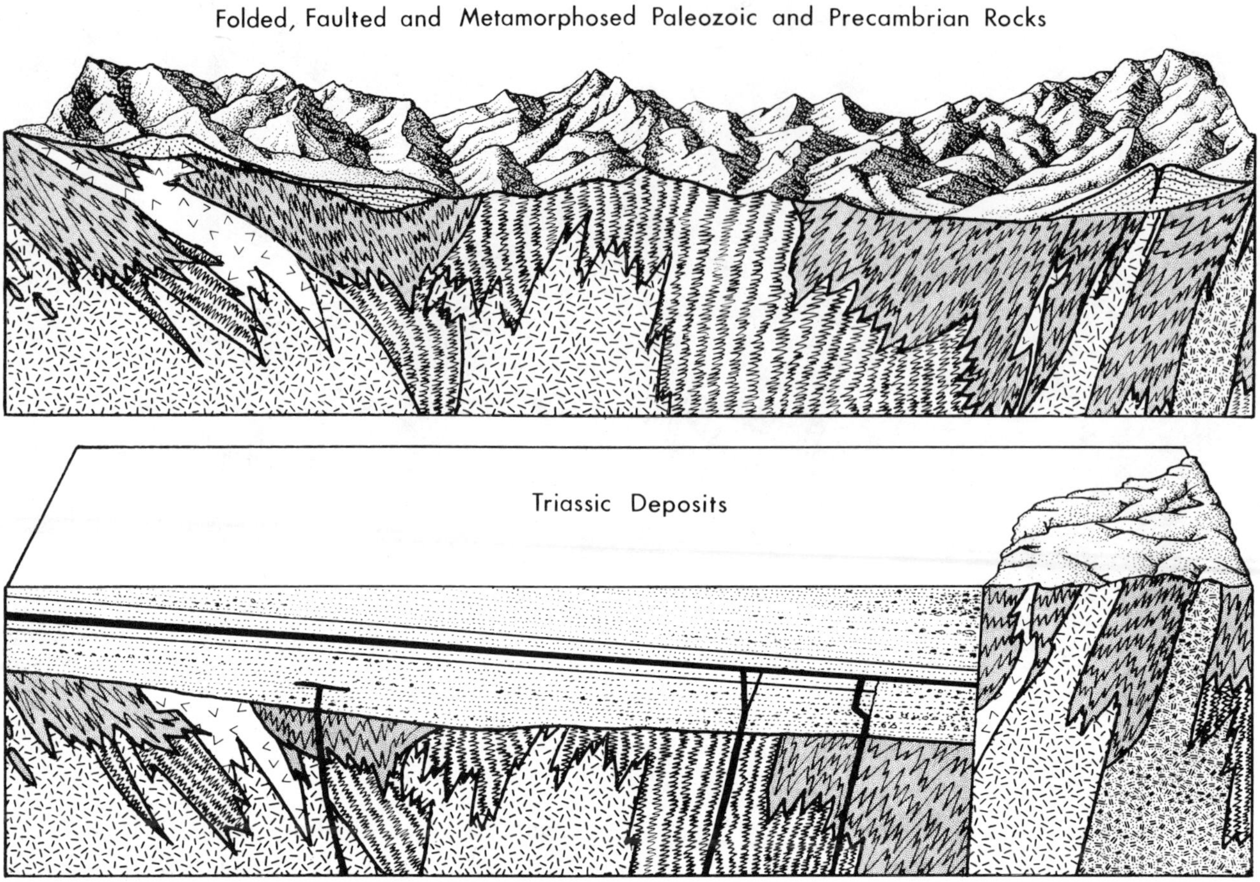

Figure 17.2. Alluvial fan sediments and igneous intrusives and extrusives in a Triassic aulacogen developed on an older continental borderland in New England.

western Pacific. The assemblages of sediments in such borderlands include normal shoreline and deltaic sediments, land-derived open shelf and deeper-water sediments, and island-arc-derived volcanic sediments in shallow to very deep water at the plate margin (Figure 17.6A). The deeper-water, volcanically derived sediments are often bedded as alternating layers of mud and coarse sand termed *flysch.* The poorly sorted volcanically derived sandstones are called *graywackes.* Some of the deeper-water sediments were deposited by very rapid underwater landslides called *turbidity flows* and are characterized by graded bedding (see Figure 8.1E and F). The land-derived sediments eventually build out over the volcanically derived sediments and are termed (in contrast to flysch) *molasse.* An idealized sedimentary sequence in a subductive borderland would start with ultramafic rocks of the oceanic crust intruded by basaltic dikes, overlain by basaltic lavas and pillow lavas, followed by black shale and interbedded black shales and graded graywackes (flysch), followed by well-sorted sandstones, skeletal carbonates, conglomerates, marine and freshwater shales, and coal beds (molasse).

Throughout the earth's history, continents have not only split apart, but they have also come back together again through the collision of their leading edges. Because the continental blocks are not easily subducted, collisions result in the compression of the rocks and sediments of the continental borderlands into folded mountain ranges and the end of subduction along the line of the range. Folded mountain ranges, such as the Urals and the Himalayas, mark the sites where ancient ocean basins were closed over and two previously separated continents became sutured. Just prior to closure, however, the trough between two colliding continents is very long and narrow and becomes filled with sediments derived from both sides (Figure 17.6B). While the subduction is still active in the narrow trough, sediments filling in the trench (especially flysch) will be continuously

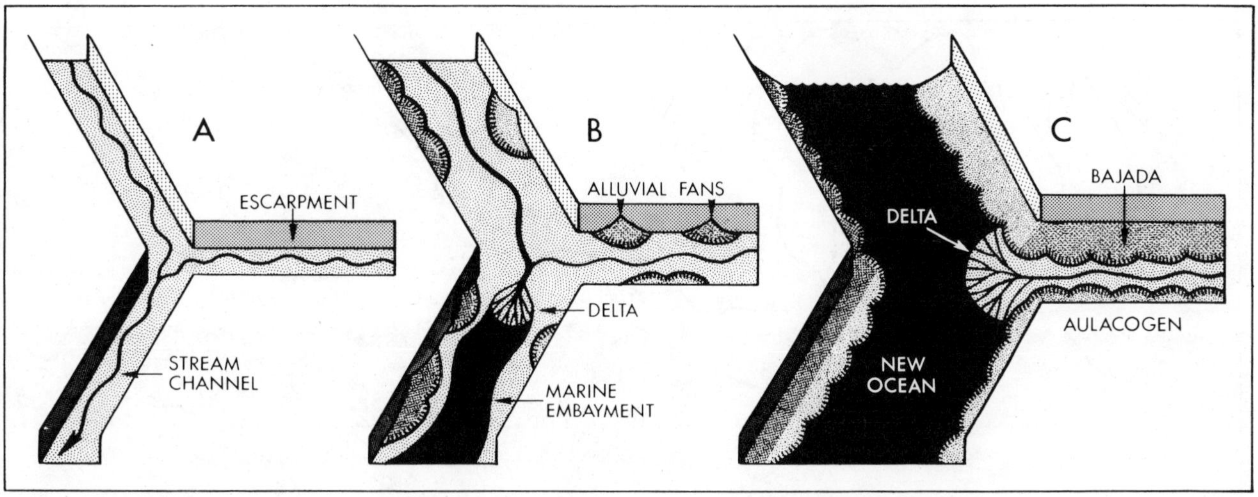

Figure 17.3. Sedimentary environments associated with continental rifting and the development of an aulacogen from a triple junction.

sinking along with the subducted plate, and sediments from the adjacent eroding island arcs and continental margins will keep the trench continuously filled up until subduction has stopped. This produces enormous thicknesses of sedimentary rocks in folded mountain belts (eight to ten times as much as in interior regions), some of which have experienced metamorphism. Compressional folding of rocks may take place even before the trough is completely filled, and erosion of igneous and metamorphic rocks produced by subduction may contribute sediment to the final stages of molasse deposition.

It is not uncommon for new rift zones to develop near older zones of collision, and a single continental borderland may have undergone many episodes of alternating rifting and collision. New rifts often develop at an oblique angle to the line of closure of an ancient ocean, so that the range of folded mountains is sliced in two and its halves eventually end up on opposite sides of the newly developed ocean basin.

Exercise

Part I

If they are available, rock samples from flysch sequences will be set out in the laboratory for you to examine. For each sample designated, record the grain size, sorting, bedding, sediment structures, and mineralogy. Make a list of criteria which could be used to distinguish flysch sediments from molasse.

Part II—Simulation of Plate Tectonics and Sedimentation

Assume a continental block that is roughly square in outline and about the size of Australia. It will be moved in an east-west direction, and you will describe its geology in different periods of the earth's history in three areas: its western margin, its eastern margin, and its continental interior. Assume initially that it is part of a much larger continent to which it is joined along its eastern margin. Draw a *sediment map* (on tracing paper) of the continental margins and the interior for the geologic time periods listed below. Use standardized colors for (1) alluvial fan sequences, (2) deltaic sequences, (3) shoreline sequences, (4) carbonate banks, (5) open marine shelf, (6) flysch areas, and (7) areas experiencing erosion.

Period 1—Rifting along the eastern margin of the continent, opening up a new ocean basin on the east. One or more aulacogens are formed on the eastern margin, and a subduction zone is formed on the western margin.

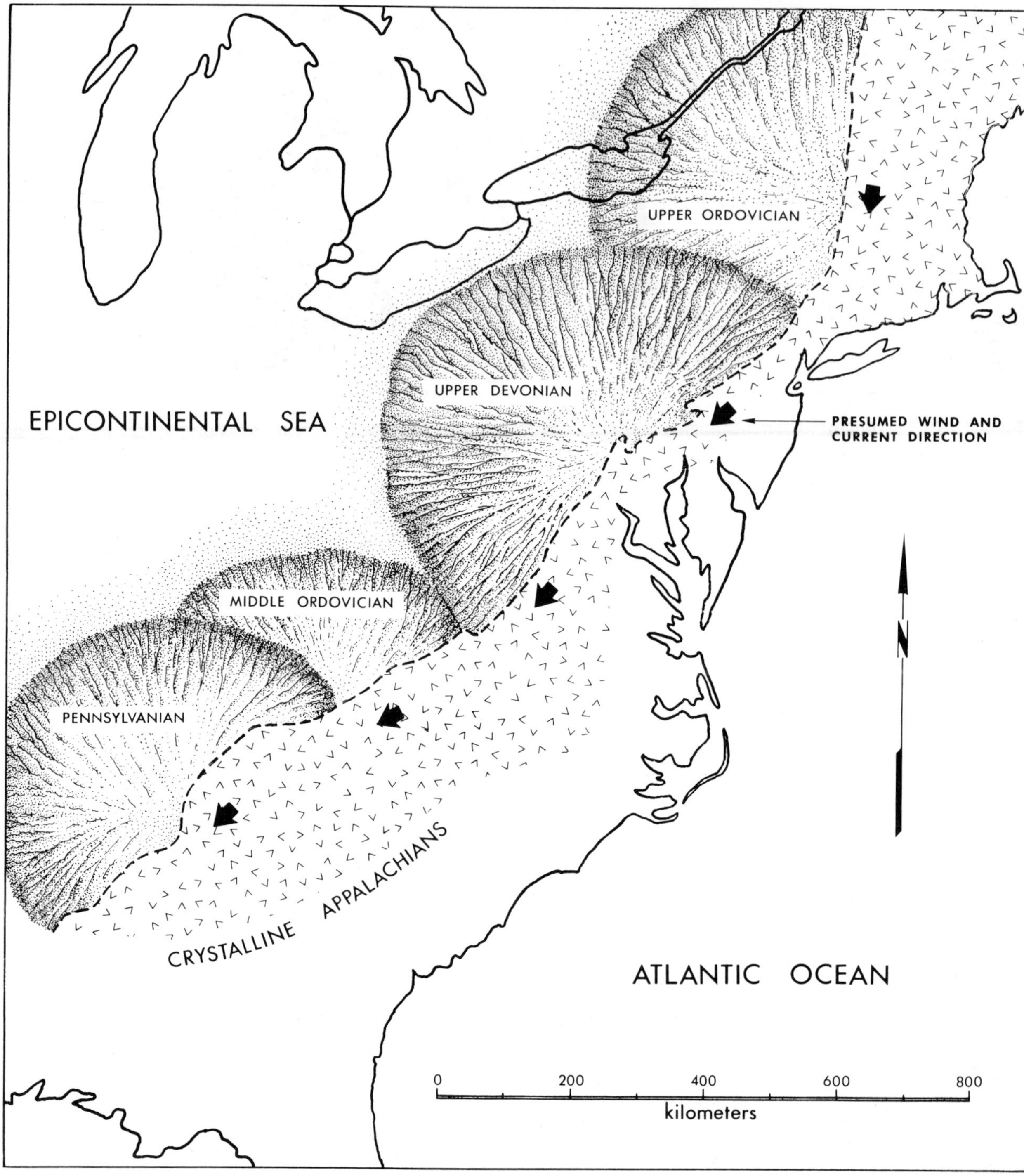

Figure 17.4. Deltaic cones developed on a continental margin that supplied detrital sediments into an interior epicontinental sea.

Period 2—Continued westward movement with high rates of spreading.

Period 3—Spreading zone in the western ocean becomes vigorous, and the continent moves eastward, developing a subduction zone along its eastern margin. Spreading rates remain high.

Period 4—Continental collision on the eastern margin and growth of a folded mountain chain.

Period 5—New rifting occurs along the eastern margin. The line of the new ocean is at about

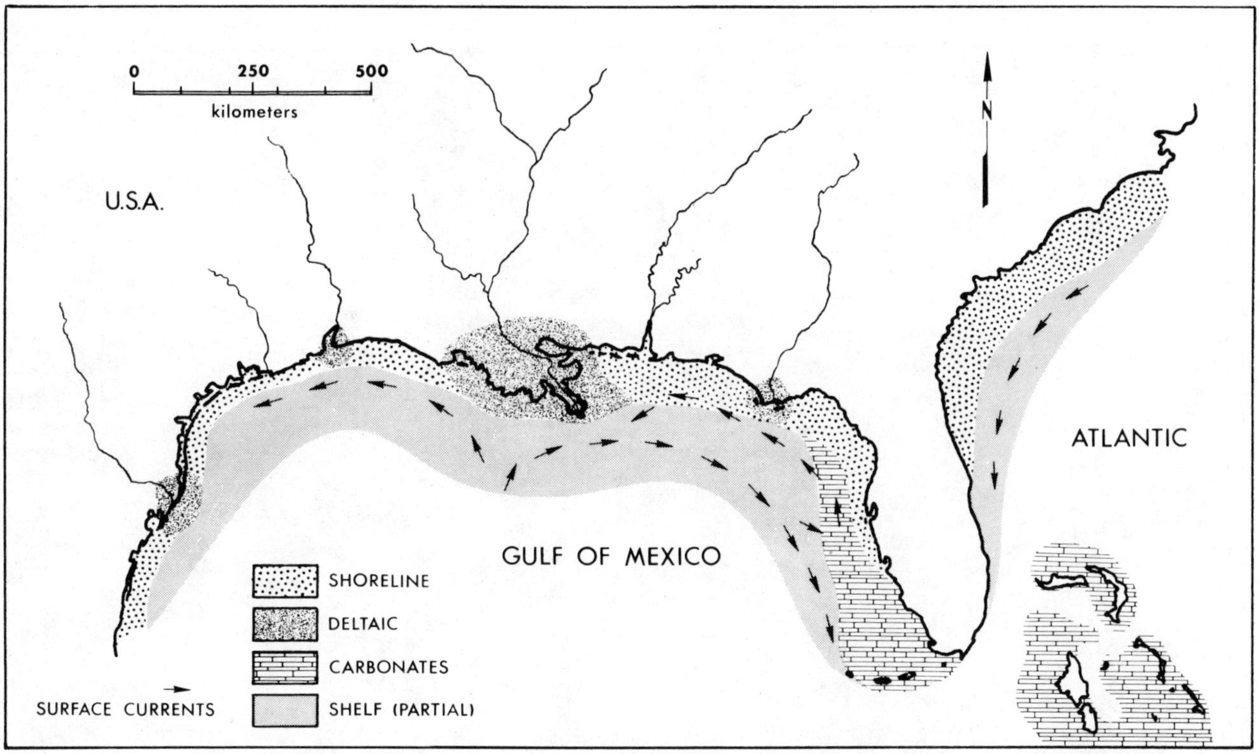

Figure 17.5. Map of the Gulf of Mexico region, showing contemporary areas of carbonate, deltaic, shoreline, and shelf sedimentation.

15° from the line of the folded mountain range, which is split in half. One or more aulacogens are formed on the eastern margin.

By superimposing the sediment maps, construct an east-west cross section through the middle of the continent, showing all sedimentary rocks and erosion surfaces for periods 1 through 5. Use extreme vertical exaggeration in showing sediment thicknesses.

Compare the history of your simulated continent to the real geologic history of North America. List the similarities and differences. Which real time periods are most like each of the five simulated intervals?

Questions for Discussion

1. What are the general relationships between time and stratigraphy in a continent-wide assemblage of sedimentary rocks? How are relationships on the continental margins different from those in the stable interior?
2. What features of sedimentation related to global tectonics could be used for worldwide time correlations of strata?
3. What effects might plate tectonics have on patterns of allopatric speciation? What ecological effects would be produced by the mixing of shelf faunas from previously separate continents brought together by plate tectonics?
4. Larger continental areas are more seasonal than smaller ones, and their shelf areas are subject to greater seasonal fluctuation in resources. With this in mind, what broad evolutionary effects on marine shelf communities would be brought about by alternating periods of continental breakup and reassembly?

Background Reading

Dott, R. H., and Batten, R. L. 1976. Evolution of the Earth, pp. 115-54, 230-41. New York: McGraw-Hill.

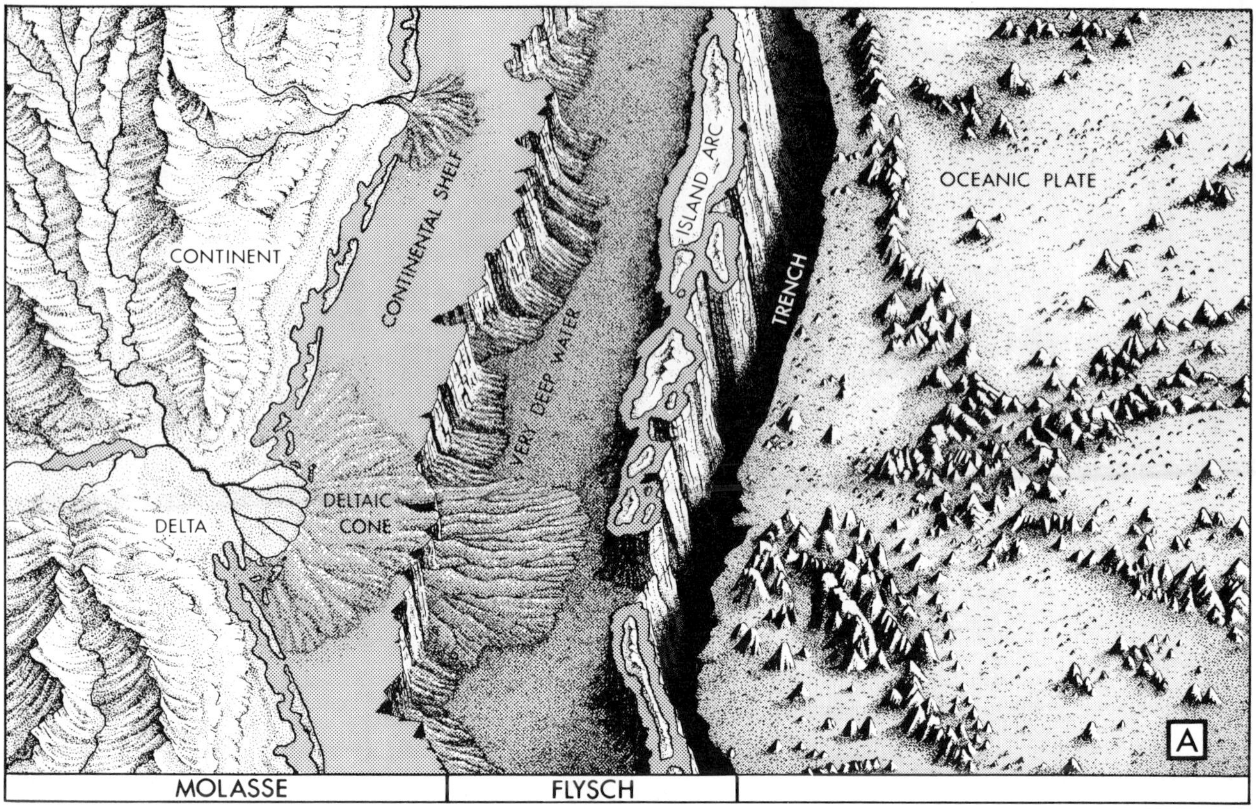

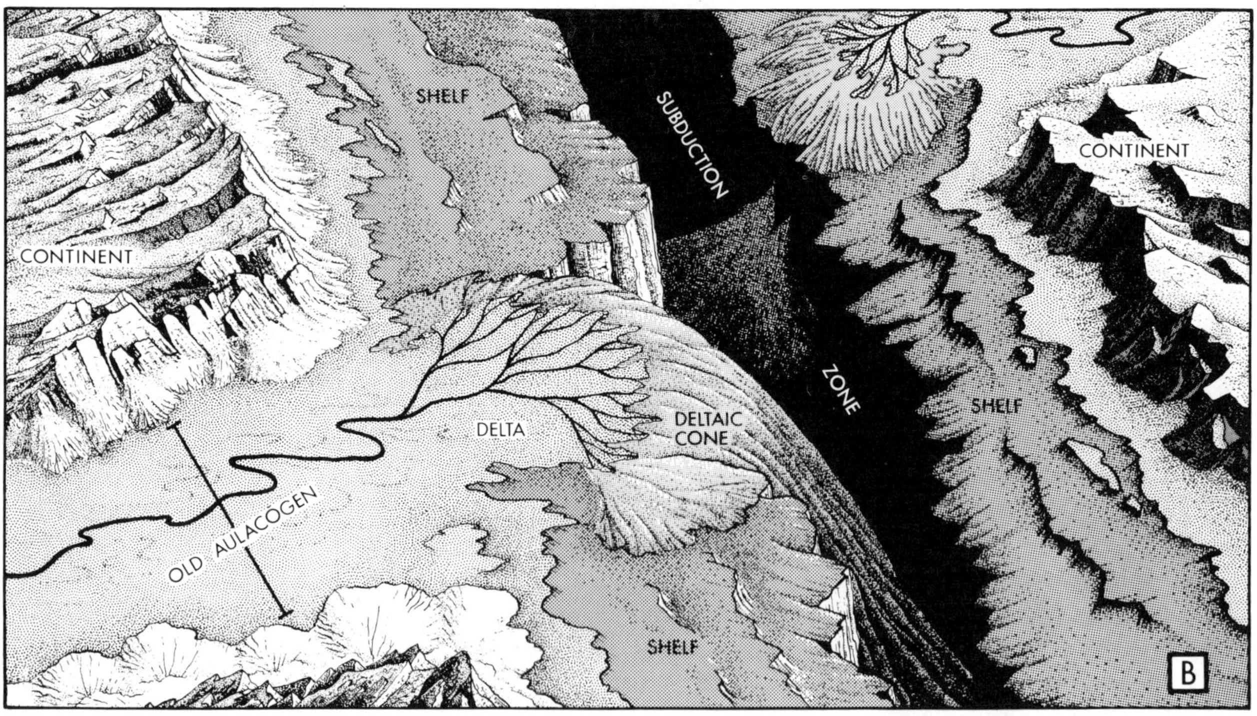

Figure 17.6. Sedimentary environments associated with subducting continental margins and continental collisions.

Sources of Illustrations

1.1 After Powers (1953) fig. 1.
1.2 After Folk (1965) p. 104, and Pettijohn, Potter, and Siever (1972) fig. A-1.
1.3 After Folk (1965) p. 107.
1.4 Based on Krumbein and Sloss (1963) figs. 4-3, 5-8.
2.1 From Hamblin and Howard (1971) fig. 131.
2.2 Original.
2.3 Partially after Mason and Folk (1958) figs. 1, 2.
2.4 Based on Fisk et al. (1954) figs. 1, 8, and Fisk (1961) fig. 4.
2.5 From Hamblin and Howard (1971) fig. 157.
3.1 After Margulis (1970) frontispiece, and McAlester (1968) p. 152.
3.2A After Fenton and Fenton (1958) p. 76.
B Based on Moore, Lalicker, and Fisher (1952) fig. 4-12.
C-F Original.
G After Beerbower (1968) fig. 10-16A.
H Original.
3.3A After Glenn (1904) pl. 100, fig. 2.
B Original.
C After Nicholson and Lydekker (1889) fig. 577.
D After Nicholson and Lydekker (1889) fig. 559.
E After Easton (1960) fig. 9.14, 1a.
F Original.
G After Shrock and Twenhofel (1953) fig. 10-41A.
H After Moore, Lalicker, and Fisher (1952) fig. 8-34, 12.
I After Martin (1904) pl. 52, fig. 5.
J Original.
3.4A Original.
B After Fenton and Fenton (1958) p. 41.
C After Moore, Lalicker, and Fisher (1952) fig. 9-24.
D After Fenton and Fenton (1958) p. 178.
E Original.
F After Easton (1960) fig. 11.17, 8a.
G After Knight (1856) fig. 2547.
H After Easton (1960) fig. 11.31, 5a.
I Original.
J After Nicholson and Lydekker (1889) fig. 782.
K-L Original.
3.5A After Easton (1960) fig. 12.2.
B After Walcott (1886) pl. 16, fig. 6.
C After Walcott (1910) pl. 33, fig. 1.
D Original.
E After Whittington (1959) fig. 353-1a.
F After Collinson (1956) pl. 8.
G After Fenton and Fenton (1958) p. 222.
H Original.
I After Clark and Twitchell (1915) pl. 14, figs. 3a, 3b.
J After Clark and Twitchell (1915) pl. 43, fig. 2a.
3.6A After Fenton and Fenton (1958) p. 97.
B After Fenton and Fenton (1958) p. 91.
C Original.
D After Ulrich (1890) pl. 63, fig. 8.
E After Moore, Lalicker, and Fisher (1952) fig. 5-4B, D.
F After Beerbower (1968) fig. 11-3D.
G After Hall (1852) pl. 40D, fig. 1a.
H-I After Moore, Lalicker, and Fisher (1952) fig. 6-10.
J After Hall and Clarke (1894) pl. 56, figs. 14, 15.
K After Hall and Clarke (1893) pl. 8, fig. 17.
L After Hall and Clarke (1894) pl. 55, fig. 5.
M Original.
3.7A After Shrock and Twenhofel (1953) fig. 14-9B, C.
B-C After Fenton and Fenton (1958) p. 128.
D After Roemer (1860) pl. III, fig. 2a.
E After Fenton and Fenton (1958) p. 128.
F After Moore, Lalicker, and Fisher (1952) fig. 18-30, 20.
G After Wachsmuth and Springer (1897) pl. 9, fig. 1.
H After Springer (1920) pl. 53, fig. 41.
I After Fenton and Fenton (1958) p. 130.
J Original.
K After Fenton and Fenton (1958) p. 131.
L Original.
M-N After Fenton and Fenton (1958) p. 130.
3.8 After House (1971) fig. 14.1.
3.9 After Hedgpeth (1966) fig. 5.
3.10-3.18 Original.
4.1 Original.
4.2 After Eicher (1968) fig. 6-1.
5.1 After Heckel (1972) fig. 9.
5.2 After Heckel (1972) fig. 3.
5.3A After Wells (1954) pl. 126, figs. 3, 5; pl. 127, fig. 4; pl. 128, fig. 3
B Original.
C After Jones (1907) figs. 155, 156.
5.4 Original.
5.5 After Bretsky (1970) fig. 30.
5.6 After Schopf (1969) fig. 1.
5.7 After Lowenstam (1967) fig. 4.
6.1 Based on Bretsky (1969).
6.2 After Lowenstam (1957) fig. 4 and Nicol (1962).
6.3 Based on Lowenstam (1950).
7.1 After Folk (1968) pp. 157, 165.
7.2 After Anstey and Fowler (1969) fig. 5.
7.3 After Folk and Robles (1964) fig. 9.
7.4A After Scoffin (1970) fig. 18.
B After Scoffin (1970) fig. 17.
C After Scoffin (1970) fig. 8.
7.5 After Shaw (1964) figs. 8-1, 8-2.
7.6 After Imbrie and Purdy (1962) fig. 13.
8.1 Based on Hills (1963) figs. 11-1, 11-2.
8.2 After Shrock (1948) figs. 48, 49.
8.3 Original.
8.4 After Harms (1969) fig. 15.
8.5 After Eicher (1968) fig. 2-12.
8.6 After Shrock (1948) fig. 207.
8.7 After Shrock (1948) fig. 148.

9.1 From Seilacher (1967b) p. 74.
9.2A From Seilacher (1967b) p. 75.
B-C Original.
D After Frey and Reineck (1973) fig. 4.
9.3 Based on Seilacher (1964, 1967a).
9.4 After Frey and Howard (1969) fig. 2.
10.1 After Gebelein (1969) fig. 14.
10.2 Based on Logan, Rezak, and Ginsburg (1964).
10.3 After Logan (1961) fig. 3.
10.4 After Logan (1961) fig. 2.
11.1 After Butler (1969) figs. 1, 2.
11.2 Based on Hsu and Siegenthaler (1969), and Shinn, Ginsburg, and Lloyd (1965) fig. 4.
11.3 After Thompson (1970) fig. 18.
11.4 Original.
12.1 From Hamblin and Howard (1975) fig. 155.
12.2 From Hamblin and Howard (1975) fig. 156.
12.3 From Hamblin and Howard (1975) fig. 149.
13.1 From Hamblin and Howard (1971) fig. 160.
13.2 After Ferm, Milici, and Eason (1972) fig. 2.
14.1A After Allen (1964) fig. 1.
B-D After Fisher et al. (1969) fig. 37 (adapted from several sources).
14.2 Original.
14.3 Original.
14.4 From Hamblin and Howard (1971) fig. 210.
14.5 After Fisher et al. (1969) fig. 13.
14.6 After Morgan (1970) fig. 8.
14.7A After Coleman and Gagliano (1964) fig. 5.
B After Coleman and Gagliano (1964) fig. 1 (adapted from Kolb and Van Lopik [1958]).
14.8-
14.11 Original
15.1 From Hamblin and Howard (1971) fig. 178.
15.2 After Strahler (1966) fig. 5.
15.3 After Strahler (1966) fig. 10.
15.4 After Dunbar and Waage (1969) fig. 18-9.
15.5-
15.8 Original.
16.1 After Smith (1966) p. 457 and Watson (1962).
16.2 After Hennig (1966) p. 61.
16.3 Original.
17.1 Based in part on Bull (1972) figs. 11, 14.
17.2 After Moore (1958) fig. 13.17.
17.3 Original.
17.4 After King (1959) fig. 34.
17.5 Based in part on Shepard et al. (1960).
17.6 Original.

Illustration Source References

Allen, J. R. L. 1964. "Sedimentation in the Modern Delta of the River Niger, West Africa." In L. M. J. U. Van Straaten, ed., Deltaic and Shallow Marine Deposits, pp. 26-34. Amsterdam: Elsevier.

Anstey, R. L., and Fowler, M. L. 1969. "Lithostratigraphy and Depositional Environment of the Eden Shale (Ordovician) in the Tri-State Area of Indiana, Kentucky, and Ohio." Journal of Geology 77:668-82.

Beerbower, J. R. 1968. Search for the Past. Englewood Cliffs, N. J.: Prentice-Hall.

Bretsky, P. W. 1969. "Central Appalachian Late Ordovician Communities." Geological Society of America, Bulletin 80:193-212.

——. 1970. Upper Ordovician Ecology of the Central Appalachians. Peabody Museum of Natural History Bulletin, vol. 34. Yale University.

Bull, W. B. 1972. "Recognition of Alluvial Fan Deposits in the Stratigraphic Record." In J. K. Rigby and W. K. Hamblin, eds., Recognition of Ancient Sedimentary Environments, pp. 63-83. Society of Economic Paleontologists and Mineralogists, Special Publication 16.

Butler, G. P. 1969. "Modern Evaporite Deposition and Geochemistry of Coexisting Brines, the Sabkha, Trucial Coast, Arabian Gulf." Journal of Sedimentary Petrology 39:70-89.

Clark, W. B., and Twitchell, M. W. 1915. The Mesozoic and Cenozoic Echinodermata of the United States. U.S. Geological Survey, Monograph 54.

Coleman, J. M., and Gagliano, S. M. 1964. "Cyclic Sedimentation in the Mississippi River Deltaic Plain." Gulf Coast Association of Geological Societies, Transactions 14:67-80.

Collinson, C. W. 1956. Guide for Beginning Fossil Hunters. Illinois State Geological Survey, Educational Series 4.

Dunbar, C. O., and Waage, K. M. 1969. Historical Geology. New York: John Wiley and Sons.

Easton, W. H. 1960. Invertebrate Paleontology. New York: Harper and Brothers.

Eicher, D. L. 1968. Geologic Time. Englewood Cliffs, N.J.: Prentice-Hall.

Fenton, C. L., and Fenton, M. A. 1958. The Fossil Book. Garden City, N.Y.: Doubleday.

Ferm, J.C., Milici, R. C., and Eason, J. E. 1972. Carboniferous Depositional Environments in the Cumberland Plateau of Southern Tennessee and Northern Alabama. Tennessee Division of Geology, Reports of Investigations 33.

Fisher, W. L., Brown, L. F., Scott, A. J., and McGowen, J. H. 1969. Delta Systems in the Exploration for Oil and Gas. Bureau of Economic Geology. University of Texas at Austin.

Fisk, H. N. 1961. "Bar Finger Sands of the Mississippi Delta." In J. A. Peterson and J. C. Osmond, eds., Geometry of Sandstone Bodies, pp. 29-52. Tulsa, Okla.: American Association of Petroleum Geologists.

——, McFarlan, E., Kolb, C. R., and Wilbert, L. J. 1954. "Sedimentary Framework of the Modern Mississippi Delta." Journal of Sedimentary Petrology 24:76-99.

Folk, R. L. 1965. Petrology of Sedimentary Rocks. Austin, Tex.: Hemphill's.

——, and Robles, R. 1964. "Carbonate Sands of Isla Perez, Alacran Reef Complex, Yucatan." Journal of Geology 72:255-92.

Frey, R. W., and Howard, J. D. 1969. "A Profile of Biogenic Sedimentary Structures in a Holocene Barrier Island-Salt Marsh Complex, Georgia." Gulf Coast Association of Geological Societies, Transactions 19:427.

Frey, R. W., and Reineck, H. E., 1973, "Holocene Sediments of the Georgia Coastal Area." In Frey, R. W., ed., The Neogene of the Georgia Coast, pp. 1-58. Athens, Ga.: University of Georgia, Department of Geology.

Gebelein, C. D. 1969. "Distribution, Morphology, and Accretion Rate of recent Subtidal Algal Stromatolites, Bermuda." Journal of Sedimentary Petrology 39:49-69.

Glenn, L. C. 1904. "Pelecypoda." In Miocene, pp. 274-401. Baltimore: Maryland Geological Survey.

Hall, J. 1852. Paleontology of New York, vol. 2. Albany, N.Y.: C. Van Benthuysen. ——, and Clarke, J. M. 1893. Paleontology of New York, vol. 8, pt. 1. Albany, N.Y.: C. Van Benthuysen.

——. 1894. Paleontology of New York, vol. 8, pt. 2. Albany, N.Y.: C. Van Benthuysen.

Hamblin, W. K., and Howard, J. D. 1971. Physical Geology Laboratory Manual, 3rd ed. Minneapolis: Burgess.

——. 1975. Exercises in Physical Geology, 4th ed. Minneapolis: Burgess.

Harms, J. C. 1967. "Hydraulic Significance of Some Sand Ripples." Geological Society of America, Bulletin 80:363-96.

Hills, E. S. 1963. Elements of Structural Geology. New York: John Wiley and Sons.

Heckel, P. H. 1972. "Recognition of Ancient Shallow Marine Environments." In J. K. Rigby and W. K. Hamblin, eds., Recognition of Ancient Sedimentary Environments, pp. 226-76. Society of Economic Paleontologists and Mineralogists, Special Publication 16.

Hedgpeth, J. W. 1966. "Marine Ecology." In R. W. Fairbridge, ed., The Encyclopedia of Oceanography, pp. 454-59. New York: Reinhold.

Hennig, W. 1966. Phylogenetic Systematics. Urbana: University of Illinois Press.

House, M. R. 1971. "Evolution and the Fossil Record." In I. G. Gass and P. J. Smith, eds., Understanding the Earth, pp. 192-211. Cambridge, Mass.: M.I.T. Press.

Hsu, K. J., and Siegenthaler, C. 1969. "Preliminary Experiments on Hydrodynamic Movement Induced by Evaporation and Their Bearing on the Dolomite Problem." Sedimentology 12:11-25.

Imbrie, J., and Purdy, E. G. 1962. "Classification of Modern Bahamian Carbonate Sediments." In W. E. Ham, ed., Classification of Carbonate Rocks, pp. 253-72. American Association of Petroleum Geologists, Memoir 1.

Jones, F. W. 1907. "On the Growth Forms and Supposed Species in Corals." Zoological Society of London, Proceedings, pp. 518-56.

King, P. B. 1959. The Evolution of North America. Princeton, N.J.: Princeton University Press.

Knight, C. 1856-58. Reptiles, Fishes, Mollusca, and Insects. Pictorial Museum of Animated Nature, vol. 2. New York: London Printing and Publishing Company.

Kolb, C. R., and Van Lopik, J. R. 1958. Geology of the Mississippi River Deltaic Plain, Southeastern Louisiana. U.S. Army Corps of Engineers, Waterways Experiment Station Technical Reports 3-483 and 3-484.

Krumbein, W. C., and Sloss, L. L. 1963. Stratigraphy and Sedimentation. San Francisco: W. H. Freeman.

Logan, B. W. 1961. "*Cryptozoon* and Associate Stromatolites from the Recent, Shark Bay, Western Australia." Journal of Geology 69:517-33.

——, Rezak, R., and Ginsburg, R. N. 1964. "Classification and Environmental Significance of Algal Stromatolites." Journal of Geology 72:68-83.

Lowenstam, H. A. 1950. "Niagaran Reefs of the Great Lakes Area." Journal of Geology 58:430-87.

——. 1957. "Niagaran Reefs in the Great Lakes Area." In Treatise on Marine Ecology and Paleoecology, vol. 2, pp. 215-48. Geological Society of America, Mem 67.

——. 1967. Adaptive Traits in Skeletal Morphology. American Geological Institute, Short Course Lecture Notes, Paleoecology, pp. HL1-HL3.

McAlester, A. L. 1968. The History of Life. Englewood Cliffs, N.J.: Prentice-Hall.

McChesney, J. H. 1868. Descriptions of Fossils from the Paleozoic Rocks of the Western United States, with Illustrations. Chicago Academy of Sciences, Transactions, Article 1, pp. 1-57.

Margulis, L. 1970. Origin of Eukaryotic Cells. New Haven, Conn.: Yale University Press.

Martin, G. C. 1904. "Gastropoda." In Miocene, pp. 131-269. Baltimore: Maryland Geological Survey.

Mason, C. C., and Folk, R. L. 1958. "Differentiation of Beach, Dune, and Aeolian Flat Environments by Size Analysis, Mustang Island, Texas." Journal of Sedimentary Petrology 28:211-16.

Moore, R. C. 1958. Introduction to Historical Geology. New York: McGraw-Hill.

——, Lalicker, C. G., and Fischer, A. G. 1952. Invertebrate Fossils. New York: McGraw-Hill.

Morgan, J. P. 1970. "Deltas—A resumé." Journal of Geological Education 18:107-17.

Nicholson, H. A., and Lydekker, R. 1889. Manual of Paleontology. Edinburgh and London: William Blackwood and Sons.

Nicol, D. 1962. "The Biotic Development of Some Niagaran Reefs—An Example of an Ecological Succession or Sere." Journal of Paleontology 36:172-76.

Pettijohn, F. J., Potter, P. E., and Siever, R. 1972. Sand and Sandstone. New York: Springer Verlag.

Powers, M. C. 1953. "A New Roundness Scale for Sedimentary Particles." Journal of Sedimentary Petrology 23:117-19.

Roemer, F. 1860. Die silurischen Fauna des westlichen Tennessee. Breslau: Eduard Trewendt.

Schopf, T. J. M. 1969. "Paleoecology of Ectoprocts (Bryozoans)." Journal of Paleontology 43:234-44.

Scoffin, T. P. 1970. "The Trapping and Binding of Subtidal Carbonate Sediments by Marine Vegetation in Bimini Lagoon, Bahamas." Journal of Sedimentary Petrology 40:249-73.

Seilacher, Adolf. 1964. "Biogenic Sedimentary Structures." In J. Imbrie and N. Newall, eds., Approaches to Paleoecology, pp. 296-316. New York: John Wiley and Sons.

——. 1967a. "Fossil Behavior." Scientific American 217:72-80.

——. 1967b. "Bathymetry of Trace Fossils." Marine Geology 5:413-28.

Shaw, A. B. 1964. Time in Stratigraphy. New York: McGraw-Hill.

Shepard, F. P., Phleger, F. B., and Van Andel, T. H. 1960. Recent Sediments of the Northwest Gulf of Mexico. Tulsa, Okla.: American Association of Petroleum Geologists.

Shinn, E. A., Lloyd, R. M., and Ginsburg, R. N. 1965. "Recent Supratidal Dolomite from Andros Island, Bahamas." In L. C. Pray and R. C. Murray, eds., Dolomitization and Limestone Diagenesis, pp. 112-23. Society of Economic Paleontologists and Mineralogists, Special Publication 13.

Shrock, R. R. 1948. Sequence in Layered Rocks. New York: McGraw-Hill.

——, and Twenhofel, W. H. 1953. Principles of Invertebrate Paleontology. New York: McGraw-Hill.

Smith, R. L. 1966. Ecology and Field Biology. New York: Harper and Row.

Springer, F. 1920. Crinoidea Flexibilia. Smithsonian Institution, Publication 2501, pp. 1-486.

Strahler, A. N. 1966. A Geologist's View of Cape Cod. Garden City, N.Y.: Natural History Press.

Thompson, A. M. 1970. "Tidal Flat Deposition and Early Dolomitization in Upper Ordovician Rocks of Southern Appalachian Valley and Ridge." Journal of Sedimentary Petrology 40:1271-86.

Ulrich, E. O. 1890. "Paleozoic Bryozoa." In Paleontology of Illinois. Illinois Geological Survey, Geology and Paleontology, vol. 8, pt. 2, sec. 6, pp. 285-688.

Wachsmuth, C., and Springer, F. 1897. North American Crinoidea Camerata. Harvard University, Museum of Comparative Zoology, vol. M20, 21, p. 10897.

Walcott, C. D. 1886. Second Contribution to the Studies on the Cambrian Faunas of North America. U.S. Geological Survey, Bulletin 30.

——. 1910. "*Olenellus* and Other Genera of Mesohacidae." In Cambrian Geology and Paleontology, no. 6. Smithsonian Miscellaneous Collection, no. 1805, vol. 53, pp. 231-422.

Watson, G. E. 1962. "Three Sibling Species of *Alectoris* Partridge." Ibis 104:353-67.

Wells, J. W. 1954. Recent Corals of the Marshall Islands. U.S. Geological Survey, Professional Paper 260-1, pt. IV, pp. 385-486.

Whittington, H. B. 1959. "Trilobita." In H. J. Harrington et al., Arthropoda 1, Part O, Treatise on Invertebrate Paleontology. Lawrence, Kans.: Geological Society of America and University of Kansas Press.

Zenger, D. H. 1972. "Dolomitization and Uniformitarianism." Journal of Geological Education 20:107-124.

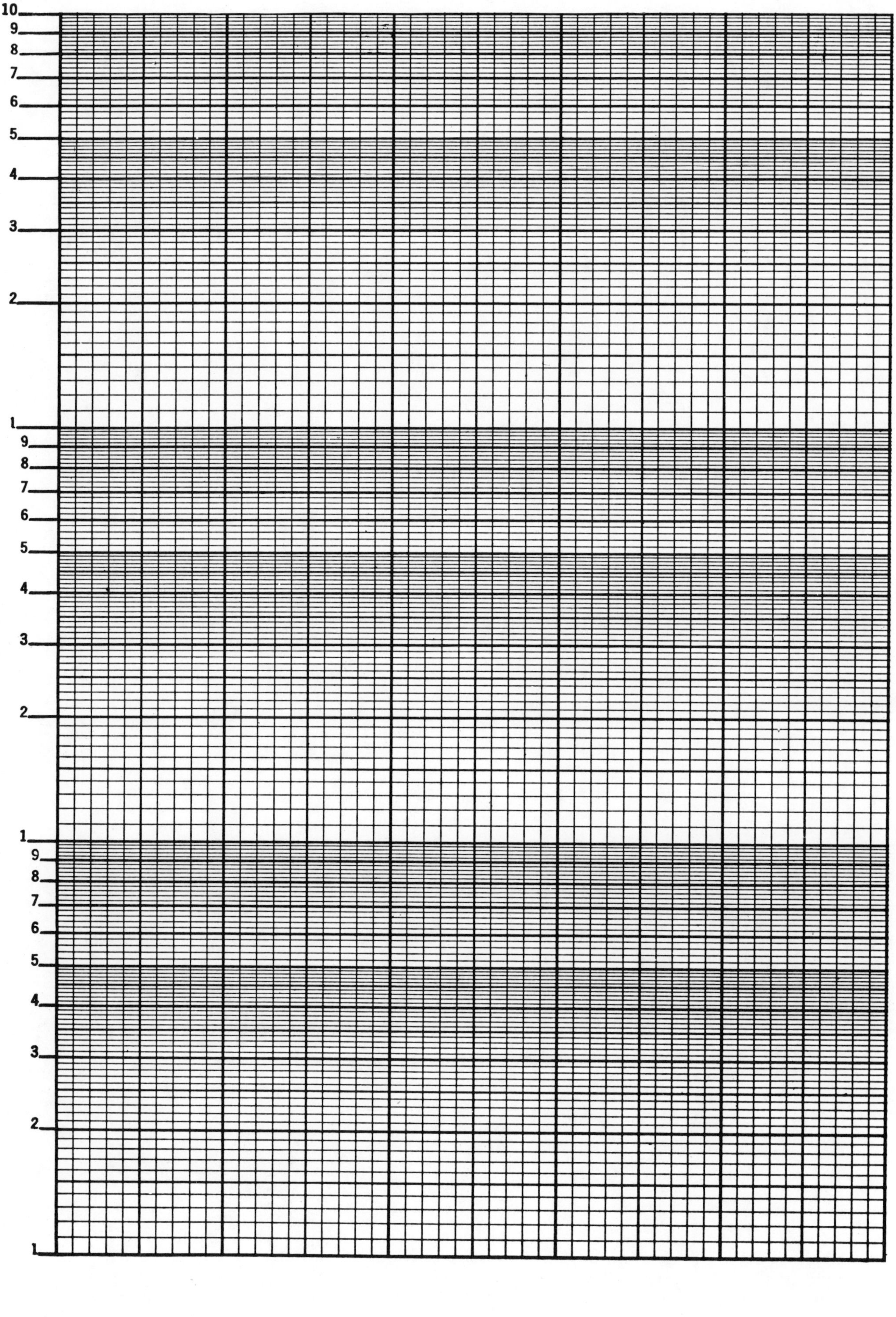
10
9
8
7
6
5
4
3
2
1
9
8
7
6
5
4
3
2
1
9
8
7
6
5
4
3
2
1